BUILT FOR SPEED

JOHN A. BYERS

BUILT FOR SPEED

A YEAR

IN THE LIFE

OF PRONGHORN

HARVARD UNIVERSITY PRESS

Cambridge, Massachusetts, and London, England · 2003

Library of Congress Cataloging-in-Publication Data

Byers, John A. (John Alexander), 1948–
   Built for speed : a year in the life of pronghorn / John A. Byers
     p. cm.
   Includes bibliographical references and index.
   ISBN 0-674-01142-2 (hardcover)
   1. Pronghorn antelope.   I. Title.

   QL737.U52B944 2003   2003041739
   599.63'9—dc21

For Karen, Anna, and Johnny

CONTENTS

Built for Speed contains the kind of information I could read forever, and is the kind of work for which I have the greatest respect: watching one animal for hours each day, days upon end, regardless of the conditions. And in a world where there are far too many middlemen and middlewomen, it pleases me to know that there is someone out there who with infinite patience stakes out antelope fawns and then by hand captures them briefly to tag and weigh them.

All literature is about loss, Shakespeare is reported to have claimed, and yet that is one of the things that makes *Built for Speed* such a wonderful book; it is literature of gain, of knowledge discovered or recovered, as if unearthed from the distant vaults of the past, through the intellectual and physical examination and excavation of Byers and other pronghorn scientists.

How tenuous is the world on which we stand; how beautiful. It astounds me in reading this book to realize what an amazing marker in time the pronghorn is: a living ghost, a victor and survivor whose every amazing quality—the twin-horned uterus, the compression of the spine, the honeycombed, ultralight bone structure, and, in the species' most discernible signature, its inimitable speed—is the legacy of evolutionary battles that were waged eons ago. (Pondering such things, one cannot help wondering if the pronghorn, like some fierce lathe, or like a god, might yet somehow carve, summon, or resurrect such splendid competitors as the North American cheetahs, hyenas, and long-legged, fast-running bears, if given another fifty thousand years—and if the landscape that birthed them remains intact.)

One of the most impressive qualities of this narrative is the seemingly nonchalant way in which Byers exercises the muscle of his scientific curiosity. Almost casually, he proposes ideas—not even hypotheses, but musings—about the various remaining mysteries of his subjects, which helps give us a workshop-level example of how a keenly observant, scientific mind operates. We are reminded that the fuel of science is curiosity. And we're shown clearly, too, the great expanse of terra incognita that lies always below, perhaps in infinite layerings. Almost never does one discovery tie things up neatly; rather it illuminates more unexplored territory and more unexamined caverns, one answer giving birth in that manner to a hundred more questions. In this regard, all scientists, I think, are pioneers.

Why do pronghorn fawns play so ecstatically, "spending" so freely their mothers' hard-earned milk? What are the evolutionary costs and benefits of such expenditure, such seeming extravagance? And why do adult males and females shove each other around so much even when there appears to be no hierarchical position to be gained and no breeding motivation? Byers does not hesitate to confess that the answers to many of these questions may lie buried in the past—while acknowledging also that they might lie hidden in the future. He is not threatened by mystery; rather, he seems nourished and stimulated by it—a healthy attitude for all of us.

In addition to being given glimpses or profiles of the working, wondering scientific mind, we're able also to witness the precision and confidence of the scientific method—the testability of hypotheses, replication, and logic. The discovery, for instance, that pronghorn are essentially red-line maxing their reproductive capabilities fits neatly and nicely into the far larger puzzle of their speedy and enduring lives. In so many ways, pronghorn are creatures literally living on the outer limits of the spectrum of physical existence, invigorating to observe and read about, and fervent testimony to the genius of landscape, and of the flares of life that burn everywhere upon a healthy, native landscape.

I love that this book, mostly about antelope, also contains natural history and observations about other inhabitants of pronghorn territory—the bison and ruddy ducks, grasshoppers, rattlesnakes, hawks, coyotes, eagles, and meadowlarks. At first I thought these fascinating accounts by the author were simply artistic, charming digressions, until I remembered: these other things are the pronghorn, as much a part of the fabric that made and sustains them as is the grassland of the Palouse Prairie itself. The symphony of ram's-horn battles, the trill of blackbirds, the tremolo of snipe, the bugling of the bull elk, all conduct the year's movement, in which the pronghorn is neither central nor peripheral, but surely vital.

Another thing I admire about this book is the scientific confidence that allows Byers-the-human-being to indulge (though that is perhaps not the right word at all; to exercise, perhaps) his emotions, unthreatened by the frequent intersection of those two paths. I am impressed by the honesty of a biologist who can write, unselfconsciously, of the reproductive struggles and caloric expenditures of female pronghorn, "Especially at this season, my heart goes out to the pronghorn. They are exposed to every kind of weather that the Range has to dish up, they are about to enter a season of slow starvation, and they will struggle to reproduce until the day they die."

And as a writer, I am delighted by Byers's observations (and prose), such as his passage describing the mid-June phenomenon in which "hundreds of meadowlark males singing in the grassland create a low flood of burbling that spreads across the prairie like the sheet of light that fireflies make at grass tops after a thunderstorm."

The spaciousness of improbably vast territory helped to create the phenomenon, the enduring blaze, of the pronghorn, and will of course be required to keep that phenomenon akindle in the coming centuries. Aldo Leopold wrote that to possess an ecological awareness is to understand that we live in a world of wounds. From this perspective *Built for Speed* is a work of literature that Shakespeare would recognize, a book about loss. So many of our wild, healthy places such as the terrain Byers examines are now but islands in a larger world of fragmentation that I believe the ecology of the future will depend upon two things: excavating, as diligently (and quickly and knowledgeably) as possible the ecology of the past, and then creating or imagining (or resurrecting) an ecology of the future, here at the surface and in-the-moment, that will reconnect this archipelago of miracles, particularly in the arid West. I'm immensely grateful to Byers for assembling so eloquently his lifetime knowledge and experience of and with this incredible creature, the pronghorn, and for demonstrating to us its place in an incredible landscape. I hope that both Byers and his subject will continue pushing on into the future, into those vast places on the map that are the terrain of country that is still wild and healthy, and that Byers will keep earning and discovering through his labors more knowledge, more science, and, paradoxically, more mystery, and bringing it to us, as he has done so well here.

The floor of the sky, the vast grassland of North America, when Lewis and Clark saw it was a magnificent ecosystem filled with millions of bison, elk, and pronghorn. The unrelenting plow and rifle pushed elk into the Rocky Mountains, and pushed bison and pronghorn to the brink of extinction. Bison and pronghorn populations are stable now, even though the animals live in isolated populations and for the most part on the fringes of their natural habitat. The National Bison Range was established in 1908 on a small mountain and some adjoining grassland in the southwest corner of the Flathead Valley in western Montana. Pronghorn were introduced to the Range in the early 1950s, and they did well on the native grassland. I began to study pronghorn on the Range in 1981. This is a book about a native North American mammal that is fleet almost beyond belief. Pronghorn are delicately built, inquisitive, and stubborn creatures with huge dark eyes and heartbreakingly long, slender limbs. Those limbs carry them over the ground with an unmatched liquid grace and blinding speed. The elegant speedsters provided me with one surprise after another for many years, and they continue to delight and amaze me. In this book, I hope to convey both my intellectual and my aesthetic appreciation of the species. Pronghorn have taught me a lot in 20 years and I hope to pass some of that learning on to you. I also hope to convey a realistic impression of what it is like to do a long-term field research project in North America. The popular media seem to promote the view that all of the interesting biology is being done on other continents, but that is simply not true.

Reader, let us carry you to the boundless plains over which the prong-horn speeds. Hurra for the prairies and the swift antelopes, as they fleet by the hunter like flashes or meteors seen but for an instant, for quickly do they pass out of sight in the undulating ground, covered with tall rank grass.

John James Audubon

1 *ANATOMY OF A SPEEDSTER*

On the short grass prairie of north-central Colorado, I observed pronghorns that had run roughly two miles put on a burst of speed and run away from a closely pursuing light plane that was traveling at 72 km. per hour.

Terry A. Vaughan, *Mammalogy*

1

Dr. James Murie, the author of the first anatomical account of pronghorn, was struck by the delicate nature of the skeleton: "no ruminant skeleton of equal size possesses such delicacy of osseous texture . . . If one were to speculate on this fact, it might be given as one reason for the extraordinary fleetness of this creature. Their rapidity of speed is related as something marvelous."[1]

Dr. Murie went on to quote John James Audubon on the speed of pronghorn, and I do so at somewhat greater length:

> The walk is a slow and somewhat pompous gait, their trot elegant and graceful, and their gallop or "run" light and inconceivably swift; they pass along, up and down hills, or along the level plain with the same apparent ease, while so rapidly do their legs perform their graceful movements in propelling their bodies over the ground, that like the spokes of a fast turning wheel we can hardly see them, but instead, observe a gauzy or film-like appearance where they should be visible.[2]

The pronghorn skeleton is not necessarily lighter than what you would predict for a hoofed mammal that weighs 110 pounds, but the bones, especially the limb bones, are long, slender, and delicate looking, as Murie noted. For example, the lower shaft of bone in the front limb is 9 inches long and just slightly thicker than a human index finger. A delicate skeleton allows pronghorn to run fast, and it also allows them to run far. Pronghorn are designed for endurance as well as for speed.

This dual design is remarkable, because a design for speed often conflicts with a design for endurance. Consider cheetahs, which are terrific sprinters, but with an almost pathetic lack of endurance—after running about a quarter of a mile with blazing speed, a cheetah is exhausted. Or consider the body types that you see at the start of an Olympic foot race. In the 100- or 200-meter dash, every contestant will have a heavily muscled torso and big thighs. At longer distances—400, 800, 1,600, 10,000 meters—the people at the start line are progressively scrawnier and longer limbed. These obvious differences reflect mechanical reality. For humans, there is an optimal body type for sprint speed and another optimal body type for distance running. These different optimal body types illustrate the adage, familiar to some biologists and to engineers, that "form reflects function."

Pronghorn run far better than any human and in fact run better than almost any animal ever to roam the earth, now or in the deep past. Pronghorn are unquestionably the fastest mammal in North America. They accelerate explosively from a standing start to quickly reach a top speed close to 60 miles per hour. Pronghorn also can cruise at 45 miles per hour for several miles. Unlike cheetahs or human racers, pronghorn are built both for sprints and for distance running.

Running speed is a consequence of two features: how fast each foot travels forward and backward, and how much distance is traversed in each forward movement of the foot. Coaches call these two features turnover rate (or stride frequency) and stride length. To run at 60 miles per hour, an animal must have very fast turnover, as well as a huge stride length. A pronghorn running fast completes a single stride in about a third of a second, which is fast, but not conspicuously faster than the stride frequencies of other ungulates of similar size. But at three strides per second, a speed of 60 miles per hour means that each stride covers *29 feet*. At this speed, a pronghorn would traverse the length of a football field in 3 and 1/2 seconds, taking just over 10 strides.

The design principles that yield both large stride length and fast turnover rate result in what I call the four-chopsticks-in-a-bratwurst body type. The chopsticks are the long, very slender and lightweight lower limbs. The bratwurst is the cylindrical body housing the upper limb bones and the muscles that move them. The upper limb segments swing back and forth rapidly because they are short and close to the body—here is the fast turnover. The long slender lower limbs turn the upper limb short-armed pendulum into a very long-armed pendulum. The distance across which the upper limb segment moves forward is multiplied enormously into a very large movement of the hoof over the ground—here is the large stride length. When a pronghorn walks, the humerus, the upper bone of the front limb, makes tiny, almost unnoticeable, front-and-back movements, but the hoof quickly whips about 2 feet across the ground. This gives the walk a stiff, almost awkward look (Audubon's term was "pompous"). When a pronghorn breaks into an easy, rocking canter (a 30 miles per hour pace that it can keep up indefinitely), the humerus swings back and forth over just a few inches with each stride. Only when a pronghorn stretches into a gallop does the humerus appear to be swinging freely, and then the hoof traverses several yards with each stride. At top speeds, pronghorn also flex the spine up and down a bit (but not nearly so much as cheetahs or most carnivores), to get some extra distance and power per stride.

The long, slender lower limb (Murie described the lower limb bone as "elegant") is the main key to pronghorn running speed but this design does carry some constraints. Pronghorn do not like to jump. They much prefer to crawl under a fence 4 feet high, if they can find a hole, rather than to jump over it. Indeed, hundreds of migrating pronghorn have been killed when heedless ranchers built fences with no bottom clearance across fall migration routes. Pronghorn even hesitate to jump across small gullies or streams. If they do jump, they usually land on the hind feet first—a goofy-looking maneuver. Pronghorn probably avoid jumping to avoid the risk of snapping the lower limb bones (9 inches long and a finger-width in diameter) when all of the body's weight comes down on them. After you've watched pronghorn for some time, white-tailed deer, when you see them run, seem slow and clumsy. But a white-tailed deer, with its shorter, heavier legs, can bound in a graceful arc over a 7-foot fence and land comfortably on its front feet.

Pronghorn have very long limbs mostly because, like all of the cloven-hoofed mammals, they have very strange hands and feet. Their hands and feet are extreme modifications of the basic tetrapod design

Profile of a pronghorn male, showing the four-chopsticks-in-a-bratwurst body type. In the front limb, the first bones that project below the torso are the paired radius-ulna. What looks like a knee is actually the wrist.

that we humans have. One of the intellectual gifts that Darwin gave us was the appreciation that staggering variety can be produced by what he called "descent with modification." The story of animal evolution contains little that is really new—true inventions are rare. Almost always, particular adaptations, such as the running-adapted limbs of pronghorn, evolve as modifications of existing structures.

The limbs of all land vertebrates are modified versions of the limb skeletons of amphibians that crawled some 400 million years ago, in the Devonian Period. During that impossibly distant age, the earliest land vertebrates established the basic limb structure that all of their descendants would have to work with. Those early amphibians, like the fish from which they descended, had 4 limbs—one pair attached to the body just behind the head, and one pair attached to the backbone near the tail. This was the tetrapod body plan, and all descendants of those amphibians—the modern amphibians, the reptiles, the birds, and the mammals—inherited its arrangement of bones.

The tetrapod plan is familiar to people because humans are mammals and hence tetrapods. The uppermost bone in the front limb is the scapula. It is held close to the ribs by muscles and ligaments, and it may also be attached to the skeleton by the clavicle and the coracoid, bones that act as struts. We humans have a stout clavicle, reflecting the climbing habits of our ancestors, and a coracoid that is reduced to a stump on the clavicle. Running specialists such as pronghorn have lost all traces of the clavicle and coracoid. In all tetrapods, the front limb below the scapula shows the same pattern: there is a single bone, the humerus, that articulates with two bones, the radius and ulna. These in turn articulate with several small cubical bones in the wrist. The wrist bones articulate with hand bones, and these with finger bones. The same pattern prevails in the tetrapod hind limb: one bone (femur), two bones (tibia and fibula), ankle bones, foot bones, toe bones. The early tetrapods established this body plan and gave it to all of their descendants. There were no subsequent inventions, only modifications. Look at your hand. You are

looking at the five-fingered pattern that those early amphibians set-
tled upon. The relatively minor, but important human modification
of the basic plan is that the innermost digit, the thumb, is rotated
about 90° so that it can oppose the other digits.

Pronghorn and other hoofed animals have done something far
more radical with their hands. To see the modifications that they
have made, try this morphing exercise. Hold your hand in front of
your face, fingers extended. Make the thumb and index finger disappear. Get rid of the associated hand bones. Do the same with the little
finger and its hand bone. You should now have a skinny hand with
two fingers. Now make the two remaining hand bones longer—*really*
longer. Make them as long as your forearm. As they get longer, make
your hand become skinnier still by fusing those two hand bones into
a single shaft. Cover the tip of each of your two fingers with a hoof.
You now have a pronghorn hand, useless for grasping (pronghorn
males in the mating season sometimes appear to lament this fact),
and too delicate for jumping, but a very effective stilt.

One year, the administrator of the National Bison Range in
western Montana, the main site where I have studied pronghorn for
the past 20 years, decided to honor a request for pronghorn from the
local Salish-Kootenai tribal wildlife officials. The tribe wanted to establish a new population of pronghorn in a valley near Hot Springs
Montana, about 30 miles to the northwest, across the Flathead River.
The tribe and the Bison Range staff captured 40 pronghorn, marked
each animal with yellow ear tags, loaded them into covered trucks,

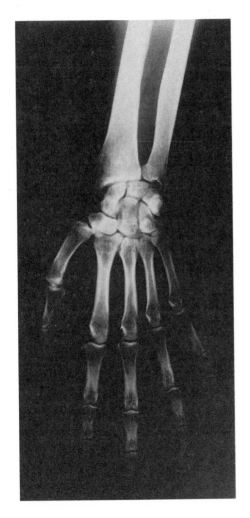

Facing page: X-ray images of the author's hand (left) and a pronghorn hand (right). In the human hand the radius (the larger of the two long bones) and the ulna articulate with wrist bones. These articulate with five hand bones. In the thumb, the hand bone is followed by two finger bones; in the other four digits, the hand bone is followed by three finger bones. In the pronghorn hand, a short portion of the bottom end of the radius where it articulates with wrist bones is visible. Below the wrist is the long, elegant cannon bone, which evolved by fusion of hand bones three and four. Below the cannon bone are three pairs of finger bones (a bit deformed, because I retrieved this leg from a carcass that was out-side all winter). The X-ray reflective objects that partly obscure the joint between the cannon bone and the first finger bone are called sesamoid bones; they act as pulleys for important ligaments.

drove them to the Hot Springs area, and released them. In that valley there are no visual landmarks that show where the Bison Range is. Nevertheless, over the next few months, most of the transplanted pronghorn returned to the Range, showing that they had some navi-gational skills that no one knew about, and that they could swim the Flathead River. They ran back and forth along the Bison Range pe-rimeter fence until let back in through a gate.

A perimeter road runs just inside the perimeter fence. One morn-ing, as I drove around a corner beyond which stretched more than a mile of straight road, I startled one of the returning male pronghorn that was still outside the fence. He started to run, and on a sudden impulse, I decided to race him. I downshifted and floored the truck to catch up. As I did, he accelerated, but I managed to pull alongside. Because he intended to cross in front of me, but could not because of the fence, we sped along for about a mile with less than 30 feet be-tween us. My attention was sharply divided by conflicting desires to watch his beautiful strides, to keep the truck on the road, and to keep an eye on the speedometer. The road was mostly flat, but on the other side of the fence that male had to negotiate many small dips. Running through this bumpy field, he maintained an even 45 miles per hour, and did not appear to be running as fast as he could. I wanted to accelerate to see if he could run faster, but an anxious croak from my assistant in the passenger seat reminded me that a sharp turn was approaching. I braked and the male shot past, over a rise and out of sight. I pulled over, a bit dazed. For a few seconds, I'd had the electrifying sensation of running with a pronghorn.

I returned to the mundane world of a slow human operating a machine, put the truck in gear, and pulled forward. Another mile down the road, the male was standing close to a gate. He wasn't even breathing hard. I opened the gate and pulled back and waited until he trotted through. The speed I'd experienced was impressive, but equally amazing was the male's endurance. He'd just run 2 miles over rough terrain at a good clip, and he wasn't winded. Other field workers have commented on the endurance of pronghorn—some report that running herds have outrun pursuing light planes. The species is built to run fast, but it is equally well built to run far.

Twenty-five years ago, Tom McKean showed some of the ways in which pronghorn are built to run far.[3] Tom studied the hearts, blood, and airway flow of pronghorn and of the domestic goat, a hoofed mammal that is not a running specialist. The comparisons were astounding: pronghorn hearts, corrected for body size, were twice as large as goat hearts. Pronghorn had 60 percent more hemoglobin, 50 percent more blood, and 33 percent more red blood cells than goats. Pronghorn airway resistance was half that of goats, because the pronghorn windpipe is about the same diameter as a large vacuum cleaner hose. Once again, these special characteristics of pronghorn are not inventions. They are just modifications of traits found in all mammals.

The features that Tom studied all are concerned with one trait that is critical for a distance runner: the rate at which the heart, lungs, and blood deliver oxygen to the running muscles. An animal's ability to run far, to work large muscles hard for many minutes, depends on this rate—that is the reason for blood doping by unscrupulous athletes.

More recently, Bud Lindstedt and his colleagues decided to take a more direct approach. They measured the rate at which a running pronghorn consumes oxygen.[4] To get this information, they used bottle-reared tame animals, and trained them to run on a treadmill. Then they trained them to wear a mask that measured airflow while they were running on the treadmill. The maximum rate of oxygen consumption that the researchers recorded was consistent with what one would predict from McKean's findings. The $\dot{V}O_2$ Max., as it is called, of pronghorn is far above the value predicted for a 100-pound land mammal. Pronghorn have a $\dot{V}O_2$ Max. essentially appropriate for a 100-pound bat, which lends authority to my impression when I raced a pronghorn that he was flying across the ground.

When pronghorn are genuinely frightened, they pack into a tight ellipse that flows rapidly and smoothly over the ground. Like a flock of starlings pursued by a hawk, or a school of fish fleeing a predator, a running pronghorn group looks like a single entity rather than what it really is, a collection of individuals. Film analysis shows that the movement is smooth because, within the group, there is 97 percent synchrony of gait (the sequence in which the four feet touch the ground: trot versus canter in horses, for example), and 92 percent synchrony of the lead foot (the first foot moved at the start of a gait cycle).[5] Pronghorn have four different high-speed gaits and a running group may shift from one gait to another. The smooth shifting while preserving synchrony of the lead foot is possible because pronghorn have the unique ability to change the fore and hind lead feet simultaneously. If you doubt that this is remarkable, get onto your hands and knees and begin to crawl. You will observe that you have a lead hand and a lead foot (knee). Now, while crawling, try to change the lead hand and lead knee at the same time. Pronghorn have some really amazing running talents.

A by-product of muscle contraction is heat—quite a bit of heat. Even mild exertion for a couple of minutes creates enough heat build-up to cause most humans to start sweating. For pronghorn, able to work the running muscles hard for long runs, the heat built up can be dangerous, especially to the brain.

Pronghorn and some other mammals are able to keep the brain cooler than the body. The heat pollution produced by exercising muscles is rapidly transferred to blood. Pronghorn cool this over-hot blood on its way to the brain with a heat exchanger. The major artery to the brain branches suddenly into a network of tiny channels. A similarly intricate network of tiny veins surrounds each channel. At the other end of the network the myriad vessels converge into a single vessel. The venous blood is cool, because it has passed through large, superficial surface areas in the nose and the ears. The cool venous blood absorbs much of the heat from the arterial network. There are two of these heat exchangers on each side. One is just below, and one is within, the eye socket.[6] The warmed venous blood continues down the neck to the heart, while the cooled arterial blood continues up to the brain. A pronghorn can keep a cool head even in the most heated situations.

For years, I watched the futile attempts of coyotes to catch adult pronghorn, remarking, but not reflecting on, the obvious unfairness of the contest. Typically, a coyote approaches and the pronghorn stop feeding to watch it. When the coyote is about 20 yards away, the pronghorn wheel and break into an easy canter. The coyote pursues at top speed but the pronghorn effortlessly glide away. The coyote usually stops after about 100 yards, tongue lolling. The coyote is completely outclassed, like a minivan racing a Ferrari.

Why is pronghorn speed incongruous? We know that adaptations, such as a long, skinny hand, or a giant heart, or a brain cooler, are created by natural selection. Natural selection occurs when individuals differ in their probabilities of death (differences in reproduction, really, but death precludes reproduction) owing to differences in a particular trait, such as the length of the hand. If individuals with the longest hands are the fastest runners, then, all else being equal, these individuals have the lowest probabilities of death and thus produce the greatest number of offspring before they die. Continue this process generation after generation after generation and you arrive at adaptations, structures sculpted by natural selection to perform a particular task. Predators pursuing prey create natural selection for running adaptations, such as a long, skinny hand. But when a predator cannot catch prey, there is no differential mortality of individuals, and no natural selection. So can a predator such as a coyote create a selection pressure that results in the prey being able to run three times faster than the predator? No.

When I eventually began to think this through, I realized that there must have been a lengthy period in the past when predators more formidable than coyotes pursued pronghorn. When I consulted information on the fossil record, I was stunned to see how completely my supposition was confirmed. About 10,000 years ago, North America experienced a large extinction event. Many terrestrial mammals disappeared. Paleontologists have advanced many hypotheses about the cause of the extinctions but at present they remain just that—hypotheses—and the cause of the extinctions remains unproven. It is certain, however, that the heartland of North America looked a lot like the Serengeti does today.[7] There were many species of pronghorn, some with four horns, some with six, some with long, spiraling, rapier-like horns. There were many other herbivores, including horses, tapirs, elephants, camels, and deer. And there were fearsome predators: lions, hyenas, cheetahs, wolves, and long-legged, fast-running bears. The North American lion was slightly larger than the African lion. The cheetahs (at least two species) were half again as large as the modern cheetah.[8] The hyena had

A scene from 12,000 years ago in North America. The American cheetah, *Miracinonyx trumani,* chases a pronghorn. (From *The Big Cats and Their Fossil Relatives,* by M. Antón and A. Turner; copyright © 1997 Columbia University Press; reproduced with the permission of the publisher.)

cheetah-like limbs (meaning that it could run very fast) and hugely powerful jaws. This, I realized, is why pronghorn can sprint at close to 60 miles per hour, and can run fast for miles without tiring or overheating. They are survivors from another world, sculpted by natural selection in that world into running machines that in today's environment blow the competition away.[9]

2 SPRING AND THE SOUNDS OF SNIPE

The male is often seen with or near the female while she is sitting, excepting towards evening or in the early part of the morning, when he mounts into the air, as if for the purpose of congratulating her by his curious song.

John James Audubon on snipe, *The Birds of America*

2   As young boys in the Eisenhower era, my older brother and I pestered our parents until they bought the family's first TV—black and white of course. Our mother soon found herself transfixed before the daily spectacle of the McCarthy hearings, but Charlie and I wanted the TV to watch *The Mickey Mouse Club*. We especially liked the show's serial story about Spin and Marty and their adventures at the Triple R Ranch. In one episode, Spin, the savvy country kid, fools Marty, the city kid, into going on a nighttime "snipe hunt." Marty spends several hours in the dark, holding his gunny sack and calling for snipe before he returns, humiliated, to camp. Some people would say that TV and the Magic Kingdom are harmless, but for years thereafter I believed that snipe were mythical, and I have met many adults who were surprised to learn that snipe are real.

Snipe and their "songs" are signs of spring in the Flathead Valley in northwestern Montana and across the northern plains. Every spring when I return to the National Bison Range, I am greeted by the still snowy peaks of the Mission Mountain range, which towers above an emerald-green grassland dotted with bright yellow clumps of arrowleaf balsamroot, rushing water, and the sound of snipe in their spring display:

woo woo woo woo woo WOO WOO WOO woo woo woo,

woo woo woo woo woo WOO WOO WOO woo woo woo

Woo gives you the general idea, but doesn't convey the texture of the sound, which varies from reedy to flutelike.

Snipe are funny-looking little birds. Mottled brown-and-white plumage hides them well in the marshy ground where they build their nests. Long toes allow them to walk in soft mud, probing with beaks more than half as long as their bodies.

On their breeding grounds in spring, male snipe set up territories and begin to display high above them. Each male's display is a set of power dives. The little bird flies high, perhaps 300 feet up, until he is no more than a dot in the sky. Then he dives steeply, sets his wings in a half-folded position, and spreads his tail feathers. The stiff outer pair of tail feathers begins to vibrate, like the reed of a clarinet, in the column of air that flows from the wings, and it is this quivering that produces the woo-woo sound. Paul Johnsgard described the sound as an "eerie tremolo." There is a crescendo as the snipe reaches the bottom of his dive and pulls upward in a shallow U. The bird then climbs rapidly to repeat the performance, diving at a rate of about eight times per minute for several minutes, before he hurtles back into the reeds. You can't see the birds unless you search the sky carefully, so their sounds seem to be part of the spring air. When several

A snipe.
(Photo by
Robert
Royse.)

males display at the same time, the air is thickened by their plaintive sounds. The sounds almost seem to come from inside your head. Snipe display when the arrowleaf balsamroot blooms, and for me, the sight of these bright yellow flowers always evokes a ghostly echo of the sounds of snipe.

Snipe flights are certainly a display of vigor and flying prowess. On a rainy evening, I once decided to count the total number of a male's dives as he flew in large circles, like a tiny feathered roller coaster car. I watched through binoculars and counted dives while a warm and pleasant drizzle fell on my upturned face. My deltoids started to tingle after 10 minutes and 82 dives, and after 15 minutes and 127 dives the tingling had become a burning that made me give up. The male was still going strong. Each spring, I find myself wondering how female snipe respond, and whether a male with better woo woos or perhaps more broadcast time attracts the best mates.

Charles Darwin, anticipating the challenge that his theory of natural selection could not explain opulent, uneconomical, seemingly disadvantageous characters, such as the long gaudy tail feathers of a peacock, proposed a force that he called sexual selection.[1] Sexual selection is a result of the importance of mating in reproduction. If you don't mate, you don't reproduce. In contrast, especially if you are a male, the more you mate, the more you reproduce. Darwin, whose intellect operated like the beam of a powerful searchlight when he turned his mind upon an aspect of natural history, realized that characters that created a male mating advantage could evolve and become exaggerated even when they created a disadvantage in all other aspects of life. A peacock's tail is the commonly used example of a sexually selected trait, but there are many other fine examples from here in North America, such as the power dives of male snipe. Many biologists study sexual selection, but there is still a lot to investigate and a lot that we don't know.

I am at the National Bison Range each spring not specifically to admire the flowering display of the Palouse Prairie, or to be quietly haunted by snipe, but to capture pronghorn fawns. I need to get my hands on each fawn for about a minute, to weigh it and to mark it in some way. There is a lot of unique and valuable information to be gleaned from birth weights and birth dates and through the ability to identify the same individuals throughout their lives. Humans collect such information on each other as a normal function of society. Getting the same information on wild animals is much more difficult. Obtaining that precious minute of personal contact with a pronghorn fawn may require several hours of stealthy observation, a careful stalk, and perhaps if I have misjudged the age of the fawn, a 200–400-yard heart-pounding chase across the prairie or down a rocky slope. The fawns are difficult to catch partly because they can outrun me even when they are 1 week old, but mainly because they hide. They hide right out in the open, and they hide very effectively.

Hoofed mammals eat plants and convert these into large masses of guts and muscle that predators find tasty. To avoid becoming a meal, hoofed mammals have evolved long legs and other specialized traits that allow them to run very fast. In the past 30 million years, the hoofed mammals on all continents have tended to become larger and faster, and only a few predators have been able to keep up in the evolutionary arms race. In the Serengeti, vast herds of wildebeest prove that most adults can outrun most attacks by predators such as lions or even wild dogs. Until the middle of the nineteenth century, the North American grassland was home to similarly vast herds of bison and of pronghorn.

Being fast works fine if you are an adult, but what if you are a newborn? All of the hoofed mammals have very fast growth rates (that's one way to shrink the window of vulnerability to predators), but still there is a period of early life when the young are small and slow and very vulnerable. Hoofed mammals have evolved two distinct ways to deal with this problem. In the "followers," such as wildebeest, the young can stand very shortly after birth, and they stay just behind or under the mother. The mother defends the young when predators approach. Most hoofed mammals, including pronghorn, are not followers, but are "hiders." Shortly after birth, at the mother's signal, the young walks away from the mother and reclines, lying motionless or nearly so for 3 or more hours until the mother returns. While the fawn is reclined, the mother remains far enough away to avoid giving up any useful information about the fawn's location, heightens her vigilance, and tries to lure approaching predators in the wrong direction.

Pronghorn fawns hide until they are about 3 weeks old. At that age, they are far too speedy for me to catch. Also, I want birth weights, so I try to catch fawns when they are between 1 and 3 days old. At these young ages they can't run very well, and they usually don't run at all—they simply flatten out and remain absolutely motionless as I approach. But finding them is not an easy task.

You might be thinking that I could just note the location of the mother, and then walk back and forth in her vicinity. The grass is only ankle-high, so it should be easy, in theory, to spot an 8–10-pound fawn lying on the ground. Actually, finding fawns in this way is extraordinarily difficult.

Mothers are within 100 yards of the hidden fawn most of the time, but this means that the careful search would be of a circle 200 yards in diameter, with an area of 31,400 square yards. Also, a mother is sometimes half a mile from her hidden fawn, and mothers are very

A 3-day-old pronghorn fawn hiding. (Reproduced with the permission of the publisher from J. A. Byers, *American Pronghorn: Social Adaptations and the Ghosts of Predators Past* [Chicago: University of Chicago Press, 1997]; copyright © 1997 by the University of Chicago Press.)

wily. If I start to search in the vicinity of a mother, she will probably casually drift away. I begin to wonder if I am searching in the right area, or if this female really has a hidden fawn at all. I start to think that a careful search of 31,400 square yards is probably a waste of time. Now that the mother has me doubting myself, and has moved out of sight so that I'm unsure of her original location, she has won the shell game. It's only a matter of time until I give up. It's much better to have a certain and fairly specific idea of where the fawn is. Then, when I search, I know that I am not wasting my time, even though the mother tries to persuade me that I am. How do I get that information on where to search? Stakeouts.

My stakeout method evolved over several years as I learned more and more about the cunning tactics and predictable habits of mothers. My knowledge of one predictable trait makes me appear godlike to freshly recruited field assistants. I know that mothers are on a schedule with the sun. A mother visits her hidden fawns at first light, about 5:00 A.M. in this part of Montana. She then goes away and returns to visit the fawns 3 hours later, at 8:00 A.M., and then 3 hours after that, at 11:00 A.M. So I have my assistants report for work at 7:00 A.M. We find a mother that I am interested in at 7:30, and I position the assistant about half a mile away. I tell the assistant to watch the mother through a telescope and binoculars. I also tell the assistant that the mother will visit the fawn in the next 40 minutes. The assistant looks at me skeptically. I tell the assistant that the fawn will suckle several times and that the mother will lick the fawn's rear end and will take the fawn's urine and feces into her mouth and will swallow them. I then say that the assistant's raison d'être as far as I am concerned comes next: the fawn begins to walk away from the mother and the mother stands and watches. The fawn walks with its head down, then stops, folds its front legs to rest its "knees" (wrists, really), then folds its back legs and sinks quickly out of sight. The assistant must carefully note the spot where the fawn has reclined and must be able to describe this location to me when I return. I return at 8:45 and the assistant looks at me with new respect: "You were right. She went to her fawn at 8:02."

I then ask the assistant to describe where the fawn is reclined. An unusually good assistant will give me an accurate description that specifies an acceptably small patch of what is usually featureless prairie. Most assistants give a description that is too vague but that they believe to be precise. Over several years, I learned how to train these assistants quickly. I give prior instructions to leave the telescope locked on the spot where the fawn is reclined. I look through the telescope, memorize landmarks along a likely route to the fawn, and also memorize several tiny details in the area close to the fawn. This odd kind of memorizing ability develops with practice. The assistant and I then drive to where we will start walking. We get out and I ask the assistant to lead me to the fawn. Of course, now that we have moved, the landscape and all the landmarks look different. As we walk out across the grassland the landmarks change even more and I see signs of increasing apprehension in the assistant, who has now realized that the "shining rock," visible as a prominent landmark from half a mile away, could now be any of 20 rocks. Most assistants now stop and with chagrin admit that they have no idea where the fawn is. When I then lead the assistant directly to the fawn my godlike status is further confirmed. But most important for me, the next time that the assistant describes a fawn's location to me will become a teachable moment. The assistant will listen to and absorb some tips on how to look at a landscape, how to recognize a foreshortening problem, and how to see and remember small details. After about a week, most assistants share my own godlike powers.

Of course, in the early years of my study, I made all of the mistakes that my new assistants make. My wife, Karen, was my only helper in the first year, and I can still vividly remember some marriage-straining moments. We'd be searching for a fawn. We'd be watching our feet, worried about rattlesnakes. The mother and maybe two or three other pronghorn females would be scolding us with their alarm snorts, and I'd be in a hurry to find the fawn and get out of there. Karen would be sure that the fawn was just to the left of that big rock, and I'd be equally sure that it was to the left of another rock 50 yards away. Eventually we'd be furiously hissing at each other and sometimes we'd drive back to the original vantage point to look for new landmarks. I will now publicly admit, 20 years later, that Karen has better spatial processing ability than I do, and that she was right about the fawn's location more often than I was. But sometimes I was right, and sometimes we were both wrong.

In those early years we also had no idea that mothers were on very distinct schedules. Many of our stakeouts thus were lengthy. If we started watching a mother at 8:30, we'd probably have to watch her until the next visit, at 11:00–11:30. Over the course of many long stakeouts we slowly gained an appreciation of what, exactly, the mothers were doing while away from their hidden fawns. Karen's first hypothesis was that the mothers were putting on a "razzle-dazzle." In other words, the mother's movements and activity, especially in the half to three-quarters of an hour before she returned to her fawns, were a kind of Chinese fire drill, designed to confuse a predator that was watching and waiting for clues about where to search. This interpretation turned out to be anthropocentric, but it is easy to understand why it was. If *you* have just sat huddled and motionless under a poncho on a 47° day with spitting rain, you also would begin to think a mother's confusing pattern of movements were designed to confuse *you*. It's easy to see yourself as a target when you are in misery. Nevertheless, like other good hypotheses, Karen's razzle-dazzle hypothesis led us into collecting the right data that eventually revealed what the mothers were doing. And what they were doing was even more interesting and more impressive than a Chinese fire drill.[2]

First, the mother gives the fawn a signal that causes it to walk away and to lie down. Thus the mother leaves no scent or visible trail to the hidden fawn. Second, as the fawn sinks out of sight, the mother stares at it and memorizes the location. Although the mother stares for only about 2 seconds as the fawn disappears, her memory of the location is perfect. While she is away from the fawn she knows *exactly* where the fawn is. Third, as soon as the fawn is down, the mother briskly moves away, often running. The mother then stays at an average distance of 75 yards from the fawn, but in attaining this average is sometimes close and sometimes far away from the fawn. Karen and I calculated that the average distance is just great enough to discourage a coyote from searching in a systematic way. Published information was sufficient to allow calculations that to my shock showed that mothers were at average distances from their fawns that made it just slightly more profitable for a coyote to hunt for ground squirrels than to search for a pronghorn fawn.

The fourth general tactic of mothers is to adjust their schedules of activity so that an observer cannot guess, on the basis of what the mother is doing, whether she has just left her fawn, is midway through a bout of feeding or reclining, or is just about to return to her fawn. The readjustment means that mothers that are within 15 minutes of a reunion with the fawn have short bouts of feeding or reclining. These are fake bouts, and their purpose is to transmit false information.

Finally, mothers anticipate their return to a fawn. In the half to three-quarters of an hour before reunion a mother usually actively searches for predators. She runs to a series of vantage points and stands stock still, her huge eyes scanning the grassland for the tiniest flicker of movement. It was this active searching for predators, combined with the fake feeding and reclining, that led to the Chinese fire drill hypothesis.

The fawns hide and the mothers do their sophisticated dance to avoid coyotes, the main fawn predator, and golden eagles, occasional fawn predators. During stakeouts, we sometimes saw coyotes find the fawn. Watching what happened was a wrenching experience, especially on the first occasion. I was watching a mother and her day-old fawn. On the previous day, I'd seen this fawn born, on a balmy bright afternoon, in a little depression decorated with lupine and balsamroot. Two hours of obvious straining and groaning by the mother, who sometimes stood and sometimes lay on her side, panting, produced a pair of dark fawns. They dried quickly in the afternoon sun. Beside the resting mother, their small heads were just visible above the tops of the grass. It was a picture of peace and contentment.

Now, on the next morning, one of the fawns was missing and the weather had changed to low clouds and a steady cold rain. "Hypothermia weather," I thought to myself, as I watched to see where the remaining fawn, picking its way through wet grass on spindly legs, would recline. Then, just at the edge of the field of view of the telescope, I saw another head, partly concealed by grass, on a low ridge top above the fawn. It was a coyote's head, I realized, as the scene began to take on that strange, slow-motion quality that tension creates. The coyote shifted its position slightly to keep the fawn in view, but also took care to stay out of sight of the mother, who was upwind. The fawn sank out of sight in the wet grass and the coyote quickly trotted closer, taking up a position from which it could observe the mother. The mother, having seen the fawn go down and having accomplished her spatial memory task, now moved quickly away. The coyote waited until she passed out of sight over a low ridge, then ran directly to the fawn and bit the back of the fawn's neck. The fawn wailed as the coyote shifted its grip. The mother, 70 yards upwind and out of sight, heard the fawn and sprinted toward it. As she did, the coyote delivered the killing shake that snapped the fawn's neck. The fawn went limp instantly and the coyote began to carry it away, the fawn's long legs trailing between the coyote's forelegs. During these 4 seconds, the mother covered the 70 yards and then struck at the coyote with her forelegs. The coyote ran and dodged, still carrying the fawn, but soon was forced to drop it to avoid being kicked.

The mother chased the coyote for 200 yards, and then ran back to the dead fawn. She stood over it for most of that day, but wandered off, calling for the fawn, at dusk. Throughout the next day and the day after that, she moved restlessly, calling for the fawn and obviously searching for it. Her anxious casting about seemed to me a physical representation of the actions of our minds when we have suffered such a loss. Her actions were grief personified. Did the mother "love" the fawn? Well, she had formed a bond to it, and that kind of bond, which exists in nearly all mammals, is the greater part of what we humans call love. Biologists still do not know exactly what happens when a bond is formed, but we do know that a bond can form with surprising speed, and that it is one of the most powerful motivators of behavior known. The fawn's carcass was of course gone—in nature, a mass of energy like that never remains intact for long.

Throughout that first spring on the Bison Range, Karen and I were surprised to see one mother after another showing up grief-stricken. Sometimes the coyotes tore the fawn's body apart. Mothers then might defend a little shredded hind leg for a day or more, before wandering off on their 2-day aimless search. The coyote mothers were acting under the direction of their own social bonds to their pups and were remarkably successful in the "fawn harvest," but I found it difficult to feel admiration for their efficiency. I saw them as marauders. They were also ruining my research. I was ear-tagging fawns so that I could follow individuals throughout their lives, but the little colored plastic ear tags were ending up as bright playthings at the mouths of coyote dens. Interestingly, the death rates of fawns that we did not capture were higher than those of fawns that we handled.[3] After we released fawns, mothers moved them to a new location; this repositioning probably erased any risk that our handling might have brought about. Still, the death rate was high. In that first summer, Karen and I ear-tagged 28 fawns; only 7 survived.

In early June, the meadowlark males, as if to offset the succession of tragic fawn deaths, begin to sing in earnest. Meadowlarks, as the name implies, are a grassland species. They build a well-concealed nest on the ground, and forage from it on a network of trails, waddling in search of insects and other invertebrate prey. An épée-like beak firmly grasps large insects, and may be probed into the ground. When a meadowlark walks, and especially when it flies, the bird reminds me of a miniature chicken. Their rapid wing beats somehow look weak, and I always have the impression that the little feathered tub of guts is about to crash. In winter, meadowlarks shift south only as far as necessary to find food.

In contrast to these secretive and ungainly habits, meadowlark males have an elegant territorial display. A male takes up an elevated perch, which may be a tall stem or a rock or, on the Bison Range, a fence post or a strand of fence wire. The male puffs out his bright yellow chest, tips his head back, opens his beak wide, and sings one of nature's truly mellifluous melodies.

The song starts with two very loud pure notes. The first is held for about a quarter of a second and the next for about half a second. Then there is a quarter-second pause, followed by three glissando-like sounds that biologists refer to as frequency sweeps. In a fre-

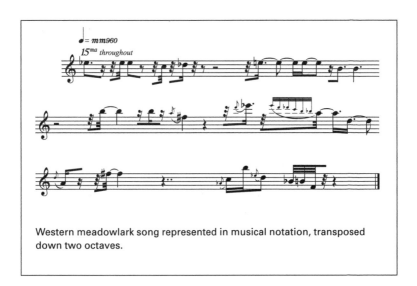

Western meadowlark song represented in musical notation, transposed down two octaves.

quency sweep, the sound passes through a wide range of frequencies in a short time. A violinist can produce a frequency sweep by bowing a string steadily while sliding the string finger from one end of the string to the other. Bats produce very loud and very big frequency sweeps, and it is probably fortunate for our sleep patterns that the sweeps occur above the range of our hearing. The three meadowlark sweeps are packed into about three-quarters of a second, and together they sound like a complex burbling. Both the purity of the introductory notes and the burble create the sensation of liquid sound. When a meadowlark male alights close by and sings, the loud clear introductory notes can be startling; they usually command all my attention for a moment. In mid-June, hundreds of meadowlark males singing in the grassland create a low flood of burbling that spreads across the prairie like the sheet of light that fireflies make at grass tops after a thunderstorm. A century ago, John Muir, writing to a woman who was concerned about the preservation of meadowlarks, had this to say:

Martinez, April, bird song month, 1895.

My dear Miss Hittell:

I heartily sympathize with you, as you know, in your efforts to save our songbirds. Better far and more reasonable it would be to burn our pianos and violins for firewood than to cook our divine midgets of songlarks for food. I am now stupidly busy writing a book and cannot do anything worth while in the way of writing for the larks. But the work of saving them by creating public opinion in their favor will have to go on year after year, and I hope I may still do something to help. I shall call the attention of the Sierra Club to the subject at the next meeting.

With best wishes,
I am very truly yours,
John Muir[4]

Toward the end of the fawn-tagging season, the only fawns left to capture are those that for the first week or so of life managed to elude both the coyotes and me. They still hide, but now, if I approach within 15 to 20 feet of them, they abruptly jump to their feet and sprint away. If I'm in reasonable physical condition, I can usually run down a 5-day-old fawn. A contest against a 7-day-old fawn is a toss-up, and a 10-day-old fawn can in effect thumb its nose at me with impunity.

To catch an old fawn, I first locate it using the stakeout method. Then, as I walk toward it, I need to be careful; if I inadvertently get too close, the fawn will jump up and run away. Older fawns usually have their heads up, and as I approach, they may be watching me. When I see the fawn and it sees me, I then must make a careful approach to the point at which I can see where the fawn's head is pointed and, more important, the direction of the fawn's body axis. Little twitching intention movements by the fawn are signs that I need to back away. Now, I need to move so that the fawn, following me with its gaze, will line up its head with its body axis. This accomplished, I instruct my assistant to remain stationary, facing the fawn. I then trot in a wide semicircle, wide enough so that the fawn will not see or hear me, until I am directly behind the fawn and about 50 yards away. The assistant uses arm signals, similar to those that ground personnel use to park airplanes, to get me lined up. At this point, I cannot see the fawn. Now I begin to walk, very carefully, toward the assistant. I must choose each step. It would ruin everything now to step on a dried balsamroot leaf. As I begin my silent-as-possible approach, the assistant begins reciting a monologue, to drown out any noise I make. The assistant also scuffs her feet and rocks from side to side, trying to claim all of the fawn's attention. As I get closer to the fawn, the assistant must speak more loudly and must perhaps take a few steps toward the fawn. I direct these changes with hand signals. The huge eyes of pronghorn are set in turrets that project from the side of the head and the field of view is large. We need to be sure that the fawn catches no flicker of movement from behind. After I can see the fawn, I closely monitor the position of its head and of its ears—an ear flicking back causes me to freeze. Now the assistant is talking very loudly and perhaps is waving her arms and I am only 6 feet behind the fawn. The tension is thick. A misstep on my part will ruin the past few hours' efforts. Somehow, I must close that 6-foot distance before the fawn sees or hears me. There is a tendency to giggle at this point, as there is when two people facing each other are carrying a refrigerator up a flight of steep stairs. Giggling is also likely if I find myself listening to the now-inane patter of the assistant's monologue. An Australian assistant sang outback songs. Another assistant talked to the fawn, telling it all about its impending capture. My most unimaginative assistant simply counted aloud. He is now a university professor.

Finally there is that moment when I must decide to commit. If I am uphill from the fawn, I can simply leap into the air and land beside it. (That is how I caught the 17-day-old monster fawn that delivered powerful kicks to my groin. Not only was I uphill, there was a stiff uphill wind, and the soaking-wet grass permitted silent footsteps.) If we are on flat ground, I simply fall forward in layout position. In either case, as I fall next to the fawn, I am reaching for a secure handhold—I will have only an instant to find one. Sometimes an old fawn turns to look at me as I am falling, and manages to get to its feet and leap away. I crash to the ground and watch the little tan rump bobbing away at high speed while ticks merrily jump onto me from the grass where I am lying. If I catch the fawn, I am always startled by the volume of its bleats. There is a sudden cacophony of noise and fumbling after 15 or 20 minutes of careful silence.

A common search route for new mothers takes me past a lovely southeast corner of the Bison Range. The perimeter road winds past and between two ponds bordered by reeds and cattails. As I drive past the first pond the water is right beside the truck. In the spring, the pond is full and its water pressure saturates the road. There are sickening sideways slips that suggest an imminent plunge. Another force that draws me off track is my enthusiastic rubbernecking at the waterfowl in the pond. You can always count on mallards and coots, but usually there are also some surprises. In the spring, I may see gadwalls, shovelers, blue-winged and cinnamon teal, redheads, canvasbacks, hooded mergansers, solitary sandpipers, American avocets, Wilson's phalaropes, and several species of grebes. If I'm lucky, I may see a sora for 1–2 seconds before it creeps back into the reeds. Anxious killdeer parents run and fly and cry in front of the truck to lure it away from their nests of speckled eggs set in small colored stones at the roadside. It is remarkably easy to stand just in front of a killdeer nest and to stare directly at the nest without seeing it. All of these bird species are a delight, but what I am most hoping to see is a male ruddy duck.

Members of the tribe of stiff-tailed ducks, ruddy ducks have short necks, stubby bodies, and as the tribe name indicates, stiff tail feathers. In this part of Montana, the ruddy ducks have recently flown in from the Pacific coast, where they spent the winter, and males have adopted their breeding colors. In the spring, a male ruddy duck has a rich dark red body, a black cap and tail feathers, a brilliant white cheek patch, and a startling electric blue bill. Like the blue of *Morpho* butterflies in the Amazon, or the blue of superb blue wrens in Australia, or the glittering blues and greens on the necks of hummingbirds, the blue of a male ruddy duck's bill is created not by a color pigment, but by a microscopic lattice that alters the way that light is reflected.

Until quite recently, ornithologists believed that this kind of structural color effect was attributable to Rayleigh scattering (named after John William Strutt, Third Baron Rayleigh, a British physicist and a 1904 Nobel laureate who showed that the sky is blue because of differential scattering of light wavelengths by gases in the atmosphere). Recently, however, Rick Prum overturned this textbook view. Rick and his colleagues showed that the blue colors of several birds, including ruddy ducks, are created by another process called coherent scattering or constructive interference.[5] The layers of collagen fibers on a ruddy duck's bill cause light to be reflected along two paths. Longer-wavelength (yellow to red) light reflects back in two paths that are completely out of phase, so these colors cancel out. Blues are reflected back on two paths in phase, so these colors are enhanced. Structural colors in animals always seem more vivid than pigment colors, and in the spring a male ruddy duck's bill almost seems to be lit from within.

When the male displays to a female, he paddles toward her with his tail feathers held vertically and his bill tucked down against his chest, which is puffed out by inflation of the tracheal air sacs. He resembles a tiny Venetian gondola. Now with short jerks of his head he begins to slap his bill against his chest right at the waterline, rapidly increasing the tempo over about 2 seconds. These slaps push air from between the breast feathers into the water, giving the slaps a flatulent quality. Bubbles form in front of the male in accompaniment to the rapid succession of hollow wet smacks. The male ends his little drum roll by stretching his neck forward to utter a soft two-syllable quack that ornithologists usually write out as "raa-anh." Sometimes, the upward movement of the bill throws a small spout of water into the air. I suppose that it is the restraint of this display that I find so appealing. Ruddy ducks are more or less the obverse of snipe—snipe have drab plumage and one of the world's most ostentatious displays; ruddy ducks have ostentatious plumage and one of the world's most low-key displays. When two male ruddy ducks display to the same female, they are so concerned about keeping their bills tucked down that physical confrontation is usually limited to bumper car–like shoving.

It's too easy to become sidetracked by life on the pond. Sometimes I "wake up," realizing that I've been watching the birds for half an hour. I have probably missed the next mother-fawn reunion.

3 *FIRST FIELD SEASON*

The leaves mingle in my memory with the leather of her shoes and gloves, and there was, I remember, some detail in her attire (perhaps a ribbon on her Scottish cap, or the pattern of her stockings) that reminded me then of the rainbow spiral in a glass marble. I still seem to be holding that wisp of iridescence, not knowing exactly where to fit it, while she runs with her hoop ever faster around me and finally dissolves among the slender shadows cast on the graveled path by the interlaced arches of its low looped fence.

Vladimir Nabokov, *Speak, Memory*

3 I met my wife, Karen, in upstate New York in the fall of 1975. One year later, we decided to spend the winter in the Baja while I was between graduate programs. I sold my impact yellow Fiat 850 spider and bought a rusty 1968 VW camper van. Bucky, as the van was soon named, carried us across the South to Tijuana and then to the tip of the Baja. The Baja highway had just been paved, and ours was the only car on the road most of the time. At Bahia de Los Angeles, we camped in Bucky on the shingle at the water's edge and were awakened by the breathing sounds of whales that echoed off cliffs across the bay. In Cabo San Lucas, then a dusty, out-of-the-way place, we put Bucky on the ferry for Puerto Vallarta. Soon thereafter, I spent two days lying in Bucky's bed, delirious with Montezuma's revenge. In Guadalajara, Bucky was by far the smallest and shabbiest of the American rigs in the RV park where we stayed. We were introduced to the expatriate American RV culture, and I learned that there was a perfect inverse correlation between the size of the RV and the size of the white poodle that it contained. In late spring, Bucky carried us up to the border at Laredo, where suspicious border agents ran a drug dog through Bucky and

pointedly asked me what I did that allowed me to spend 3 months in Mexico.

Safely across the border, we drove home across the South, stopping in small towns and asking the locals for advice on where to eat. At our respective parent's houses we loaded Bucky with furniture and other possessions and then drove to Chicago, where I would spend a second research summer at the Brookfield Zoo. Ben Beck was the creator of the summer research program. That summer, we lived in Ben's basement, and I spread a large quantity of motor oil over the Beck's driveway in the process of rebuilding Bucky's engine.

In August, we headed west to Colorado, where I entered graduate school. A year later, Bucky took me to my field site in Arizona, where I studied collared peccaries, 40-pound piglike animals that occur from Arizona to Argentina. The site was rough and I abused Bucky unmercifully for a year, probably in retribution for his refusal to produce any heat during my winter drive across the Colorado Plateau.

In the spring of 1980, Bucky made one more trip to the East Coast, where we collected more possessions, and then, heavily laden, took us back to Colorado. Here Bucky was ignominiously hitched behind a rental truck and towed to Moscow, Idaho, where I was to join the Biology Department at the University of Idaho.

And so on a cold May night in 1981, after my first academic year as an assistant professor, I found myself again huddled under a quilt on Bucky's comfortable bed. We were parked beside the machine shop at the Bison Range. I had awakened at 2:00 A.M. to the sound of rain on the roof, and I was dreamily recalling the many nights I had spent in Bucky. I was somewhere in the Baja when I began to realize that the sound of rain on the roof was different, somehow. Crap! The mattress that my wife and I would be using that summer was rolled up on the roof like a giant éclair, soaking up water. I clambered outside with a tarp and threw it over the mattress. The wet grass was icy on my bare feet and my breath made clouds. I figured that some pronghorn fawns had been born already and I briefly wondered what this night was like for them, before I slid my cold feet back into bed.

On the next morning, we woke to low clouds and continuing rain. We folded up the bed, rescued the sodden mattress from the roof, and stuffed it into the back of the van. It would have to dry out later. Our top priority was to find summer housing.

We drove to the first of two prospects that Bison Range staff had found for us. It was a basement apartment in Dixon. Dixon was (and still is) a shabby remnant of a once vigorous railroad town. One paved street ran through the three blocks of downtown. At the edge of the street, the pavement faded out in crackling hunks into dirt and the colorful round stones that are found everywhere in the Flathead Valley. Downtown had a post office, a general store with an ancient gas pump in front, and a seedy bar. Washboard dirt streets stretched uphill for two or three blocks above downtown. Old trailers were on some lots, houses on others. Horses stood listlessly in a few yards, some mean-looking dogs in others. Rain plinked off of upturned steel barrels and old car bodies rusted to the color of tobacco juice. Over the phone, the owner had given us directions: two blocks up from the post office, turn right at the green trailer, then left at the old logging truck. The owner was waiting for us in his car. We walked across a small yard choked with knapweed and the owner opened the door —no lock.

The owner led us downstairs to the rental unit. Halfway down the stairs, we were assailed by a powerful stench compounded of mold, stale beer, grease, and cigarette smoke. I knew that Karen has a very sensitive nose, so I glanced back at her. She had discreetly placed a hand over her nose and mouth. This underground apartment had no windows. Two glaring overhead lights were reflected on the flat greasy carpet. There was a living room with a kitchen at one end, and a small very dark bedroom, which seemed to be the main source of the mold smell. On the bathroom wall there were three smudges in a horizontal line about three feet above the toilet, showing where the previous tenant had rested his forehead and hands on the wall when he urinated. Next to the kitchen was what looked like a bar. The owner pulled a string and a red fluorescent light hummed and flickered luridly, illuminating a tiny sink behind the bar. "I put in this wet bar myself," he said.

Back aboveground the rain smelled wonderful. Beyond the old train terminus and through ragged low clouds, we could see a turquoise line that was the Flathead River. The steep hills above the river were only partly visible. The wind was beginning to pick up and to the west several startling blue holes appeared in the clouds. By the time we had driven the 5 miles back to the Bison Range gate, and then another 5 miles into the Moiese Valley, where our second prospect waited, the rain had stopped and the sky was half blue.

The second place was a new house with big windows, set on a dry knoll with no visible neighbors. There was a garden plot and a deck that faced the mountains. We paid $125 for the first month's rent and unloaded Bucky. We set the wet mattress inside to dry.

An hour later, we left for the Bison Range to look for pronghorn fawns. The clouds had closed in again and steady rain fell. We spotted a mother with two young fawns and waited until the fawns reclined. Then, in the oddly quiet late afternoon of a sodden prairie, we had our first lesson in noting a position from half a mile away well enough to allow us to walk to it. We searched, with increasing incredulity, for half an hour before we found the fawns. By that time the mother and four other females had approached us with flared rump patches and scolded us with their alarm snorts. The rump is covered with long white hairs; when pronghorn erect these to signal alarm, the rump becomes brighter and appears to double in size, and an alarm scent, described as similar to that of buttered popcorn, is released.[1] The animals signal a higher level of alarm with a loud sneezelike snort. I was surprised that the fawns, though soaked to the skin, felt hot when I held them. We noted the sexes, attached our first set of ear tags, and released the fawns. They ran away on spindly legs and then the mother dashed in front of them to lead them away.

We walked back to Bucky. We were now as soaking wet as the fawns had been, but not nearly as warm. Karen had blue lips and was shivering. When I turned the key, Bucky wouldn't start. After some cursing and fumbling around with stiff fingers, I realized that the fault was in the switch. I rigged a hot wire and we drove to Allentown, and took a room in the only local motel, across Highway 93 from the Ninepipe Wildlife Refuge. We took long hot showers. My record of summer expenses in the back of my 1981 field notebook reads, "Overnight in Allentown motel—$17.00." Our first field season had begun.

Fifteen thousand years ago, a glacier, the Flathead Lobe of the great Cordilleran ice sheet, filled much of the Flathead Valley. The glacier rode over the northern part of the Mission Mountains, grinding their tops smooth. To the south, where the glacier thinned out, it ground against the mountains and its tributary glaciers flowed off the peaks, cutting them into the sharp peaks and ridges that tower over the Bison Range. To the west, the Purcell Trench Lobe dammed the Clark Fork River, forming glacial Lake Missoula, which filled the southern Flathead Valley, the Bitterroot Valley to the south, and the canyon of the Clark Fork River almost all the way to Spokane. Lake Missoula was vast and almost 2,000 feet deep, but its depth varied as the glaciers slowly pulled back. Also, about 40 times in its history, glacial Lake Missoula abruptly drained across eastern Washington when the Clark Fork ice dam was breached. (Because ice floats in liquid water, it is a poor material for a permanent dam: water begins to seep under the dam, and after a through channel is made, complete collapse is swift). These epic floods, called Jökulhlaups by geologists (jökull = glacier, hlaup = floodburst), ripped off all of the soil, gouged out massive chunks of the underlying basalt, and created a strange area of potholes and irregular drainages in eastern Washington called the channeled scablands. In places, 100-foot ripple marks

The horizontal lines on Antelope Ridge and on the mountain behind mark the shorelines of glacial Lake Missoula.

still persist, and from space, the scablands look like a giant braided stream. Geologists estimate that the maximum rate of flow during floods was 386 million cubic feet per second—about 10 times the combined output of all current rivers in the world.

Most of the Bison Range was under water when Lake Missoula

filled the valley, and over centuries as the lake drained and refilled, a series of shorelines were cut. Today these appear as a dozen or so horizontal lines that run across the upper slopes of the Range. In our first field season, I often gazed at those upper slopes (I still do), and I saw the horizontal lines, but I did not really perceive them. I did not recognize that they were a curious feature that demanded an explanation. Then, after someone pointed out what they were, they stood out, as an obvious beacon from the past. In his autobiography, Charles Darwin described the same experience:

Next morning we started for Llangollen, Conway, Bangor, and Capel Curig. This tour was of decided use in teaching me a little how to make out the geology of a country. Sedgwick often sent me on a line parallel to his, telling me to bring back specimens of the rocks and to mark the stratification on a map. I have little doubt that he did this for my good, as I was too ignorant to have aided him. On this tour I had a striking instance how easy it is to overlook phenomena, however conspicuous, before they have been observed by anyone. We spent many hours at Cwm Idwal, examining all the rocks with extreme care, as Sedgwick was anxious to find fossils in them; but neither of us saw a trace of the wonderful glacial phenomena all around us; we did not notice the plainly scored rocks, the perched boulders, the lateral and terminal moraines. Yet these phenomena are so conspicuous that, as I declared in a paper published in the *Philosophical Magazine,* a house burnt down by fire did not tell its story more plainly than did this valley.[2]

Often, signs of the past are obvious, and all around us, but we cannot see them until our minds are prepared to do so.

The Flathead glacier heaved up a lateral moraine of rubble against the mountains, and when it retreated, left a large terminal moraine, which is now the long rise that Highway 93 ascends as it makes its straight course up the valley toward Kalispell and Glacier National Park. As the glacier retreated to what is now the south end of Flathead Lake, it left a huge bed of rubble, studded with blocks of ice, in the valley floor. The ice blocks melted and sank into the rubble, creating thousands of potholes. Today the potholes and the plants that grow in and around them are important habitats for breeding and migratory waterfowl and other birds. Land encompassing some of the potholes as well as a large dam-augmented lake forms the Ninepipe National Wildlife Refuge. The Bison Range also is a wildlife refuge, and the two are administered jointly from the Bison Range office. Wetlands such as Ninepipe once were common across the northern plains, but most have been drained to create farmland. The national wildlife refuges preserve isolated bits of these wetlands, and they are increasingly the only habitats available to many bird species, especially during migratory rest stops.

Ninepipe, in case you had any doubt, is pronounced 9-pipe. We did not know that when we first arrived. We did know, however, that Ninepipe, the Bison Range, and most of the southern Flathead Valley were within the Salish-Kootenai Reservation. Thus one of us (I prefer not to remember who), reading the wildlife refuge sign for the first time, said "Nin-ē-pip-ē." And for about 2 weeks thereafter, we both told Bison Range staff and other researchers and Kay, our land-lady, about interesting birds that we had seen at Nin-ē-pip-ē. Finally, someone drew us aside and said, "You know, it's 9-pipe." The name honors a man named Ninepipes for remarkable peacemaking efforts in which he smoked the pipe with nine other tribes.

Meanwhile, the summer began to acquire a steady rhythm. We rose about 4:30 A.M. and arrived at the Range about 5:30, when the eastern sky became bright. Most of the 28 fawns that we had ear-tagged as well as most of the untagged fawns were dead (killed by coyotes and eagles), but a group of 12 tagged and untagged fawns and their mothers persisted in a pasture adjacent to the visitor center that was under construction. The tour road ascended a ridge above the pasture, allowing us to park Bucky at a favorable vantage point, open his sliding door, and observe the fawns and mothers through a telescope. This arrangement allowed us to keep the fawns in view, except when the group disappeared down a deep draw that we soon named "the whale," because it swallowed whole groups without a trace.

One morning in early August, when we were parked at our usual vantage point, a Fish and Wildlife Service truck pulled up behind us. The assistant manager had driven up to tell us that there was a mes-sage from the Department of Biological Sciences at the University of Idaho. I was to call. I knew what the call was about. When I left in the spring, the day after turning in final grades, the university was going through the ugly process of declaring a financial emergency to pave the way to firing faculty. I knew that there was a general last hired–first fired policy, and I was the newest faculty member. So that afternoon, Karen and I drove over to Allentown, where there was an outdoor pay phone (no phone at our house in the Moiese Valley). The department chairman answered and came right to the point as, I later discovered, he always does: "We had a department vote on whether tenure should be the basis for making financial emergency reductions"; he paused, then went on, "and we voted that tenure would not be the basis." So the call had been to let me know the good news—my new colleagues had voted to keep me even though that would mean embroiling the department in legal wrangles when a tenured member was fired.

At the time, standing across from Ninepipe and squinting into the sinking western sun while a flight of ducks set their wings to land on the lake, I found it difficult to feel the appropriate gratitude. The university and its machinations seemed very far away and somehow irrelevant. Such is the intoxicating nature of fieldwork.

Toward the end of the first summer, Karen made a casual remark that changed the course of the pronghorn project for many years. She mentioned that it seemed as if the social rank of each fawn corresponded to its place in the birth order of all fawns in the group.

Social rank means the same thing as dominance status, or dominance rank. Thousands of animal species, including humans, live in societies that are shaped by dominance. A dominant individual is able to change the behavior of a subordinate, usually when there is a contested resource such as food, a resting place, or the opportunity to mate, using gestures alone. Sometimes, the gestures are subtle to the point of being invisible to the casual observer—a quick glance or a short stare is all that is needed.

Pronghorn are a bit more obvious in their dominance interactions. Dominant individuals give a pointed stare, or lower the head and make a butting motion at the subordinate. The subordinate signals its assent to be dominated by turning or moving away. There are three kinds of dominance interactions between females. In what I call a feeding displacement, one animal forces another to relinquish a feeding site. The dominant individual walks quickly and directly at the feeding subordinate, holding her head up and forward in a genuinely snooty way. If the subordinate does not immediately move away, the dominant lowers her head and butts at the subordinate. Perhaps the dominant appropriates succulent plants that the subordinate has found, but my impression is that the main point is simply to make it known who is the boss. In a bedding displacement, a dominant female makes a snooty approach toward a reclined female, who is forced to get up and move away. The dominant female then usually reclines on the vacated spot, with a smug, self-satisfied attitude.

A simple displacement among females looks just like the other two kinds of interaction except that no resource (feeding site, bedding site) is involved. Here it is obvious that the action of the dominant individual has no purpose other than to assert status. Actually, it seems that little of tangible importance is at stake in any dominance encounter. The conflict seems to be over who is dominant rather than over who may claim a spot to feed, recline, or stand. In adult society, each pronghorn female experiences about 20–30 dominance interactions with other females per day. This level of conflict is higher than that of any other hoofed mammal except the ultra-violent mountain goat.[3]

In the fawn social groups that form after mothers bring their surviving young together at the end of the hiding period, each fawn experiences dominance interactions at a much higher rate—about 80 to 100 per day. Also, while the fawns are sorting out their dominance relations, the interactions are much more violent. A dominant fawn approaches a subordinate fawn, lowers its head, lunges forward, and slams the other fawn in the ribs. The subordinate moves away and the dominant pursues, butting the other in the rump repeatedly. An adjacent fawn, viewing the interaction, may lunge in and butt the subordinate in the side. Other fawns may run from some distance away to join in the enforced humiliation. The mother of the target fawn may try to intervene, but unless she is dominant to the mothers of the bullies, they will thwart her. So mothers intervene rarely. Far better, it seems, to endure the spectacle of your own fawn's humiliation than to endure humiliation yourself.

Throughout our first field season, the data that Karen and I collected began to put together a clear picture of what I have just described. Fawns, if they were not resting, playing, or trying desperately to suckle, were busy vying for dominance. By the end of the summer, dominance relations among fawns were stable and obvious. There was a clear hierarchy of dominance status among male fawns, and an equally clear hierarchy among female fawns. Karen then made the offhand but disarmingly astute observation that characterizes her approach: She said that it seemed to her that the dominance status of fawns was linked to their birth order. When I checked the

A few seconds in the never-ending dominance struggles between females.
Top: The female with her rump to us forces the most distant female to get up.
Facing page, top: The female forced up turns toward her displacer, who is
forced to walk away; the reclined female on the left watches these proceedings.
Facing page, bottom: The female forced up now tries to displace the watcher,
who stands up, but threatens the other female, who is forced to turn away.
The most dominant female of these four remains placidly reclined throughout
the interaction.

data, Karen's idea was confirmed. Within each sex, there was an almost perfect correspondence between birth date and dominance rank. The earliest-born fawn was dominant to all fawns. The second-born fawn was dominant to all fawns except the first-born, and so on. The last-born fawn (GY—green tag in the right ear, yellow tag in the left) was subordinate to all fawns.

How did the birth-order hierarchy arise? The mechanism was mindlessly simple. Pronghorn fawns grow fast. Birth weight is about 8 pounds, and each fawn gains about $\frac{1}{2}$ pound *per day*. When fawn groups first form, a difference in age of 5 days between two fawns means that the older is $2\frac{1}{2}$ pounds, or about 32 percent, heavier than the younger; these size differences are easy to see. The bigger fawns take advantage of their size to dominate the smaller fawns. By the end of the summer, the size differences among fawns are still apparent, but the percentage differences are much smaller. But by this time, the dominant fawns have done such an effective brainwashing job that the subordinates rarely, if ever, challenge the status quo. I wondered what would happen to the status quo in the future, and I studied that question for many years.

4 *THE ADULT BULLIES*

Nobody who has not been in the interior of a family can say what the difficulties of any individual of that family may be.

Jane Austen, *Pride and Prejudice*

4

Three o'clock in the afternoon. Light in sharp yellow bars ripples across the ceiling as faint puffs of wind move the venetian blinds. From the eaves, the eerie calls of starlings that for the moment are imitating red-tailed hawks merge with a dream about a huge insect that beats its wings furiously against the walls of a collecting jar. Voices. The click of a horse's feet in the driveway. The strange calls of the starlings cause these perceptions to slide and melt like a Dali watch. Images of friends from college appear and then a deep sadness washes over me. I sit up, wet with sweat, and try to remember where I am. Those damn starlings.[1]

The voices outside become clearer. Nich Hall and Jake Matson, my field assistants, are talking to Steve McCready, the owner of the farmhouse that Karen and I rent, about the Bob Marshall Wilderness. I stand up, still groggy and disoriented, and separate two blind slats. Hot yellow light washes over everything and glints off the backs of the horses. Jake and Nich have their hands up, shielding their eyes. To the west, a single big thunderhead sits over the Little Bitterroot Range. In half an hour we'll get into the truck to drive over to the Bison Range for the afternoon/evening data session. I step out onto the back deck, which faces east, toward the Mission Mountains. They are hazy in the full afternoon light. Only a few patches of snow cling to the highest peaks. I turn on the outside faucet and hold the hose over my head. The water is hot, then icy. I gasp and now I am fully awake, and beginning to think about which fawn groups to try for first.

The Bison Range is at 47.2 degrees north latitude. As the summer solstice approaches, days become very long. To the east, the almost vertical wall of the Mission Range is backlit with deep blue light at 4:30 A.M. and to the west, the sky is not black until after 11:00 P.M. During the first 5 years of the study, when I concentrated my efforts on describing the development of behavior of many individual pronghorn fawns, the long days imposed a strange schedule. I had to be on the Range by 5:00 A.M., when full daylight arrived and the fawns began the first and biggest daily activity bout. From about 11:00 A.M. to 6:00 P.M., during the heat of the day, the fawns spent nearly all of their time reclined, or standing rather listlessly and nibbling vegetation. Another, smaller activity peak occurred between 6:00 and 10:00 P.M. After I learned that most social behavior in a fawn group occurred at these dawn and dusk periods, I concentrated my observations on these times and, like the fawns, took a siesta in the middle of the day.

From the side deck of our house, I could often use binoculars to spot groups of pronghorn in Alexander Basin on the Bison Range, about 3 miles away. This was a way to reduce search time. On the Range, once my assistants and I had found a group that contained fawns that we needed to watch, we'd position the truck close to it, and set up a telescope on a tripod pointing in the right direction. Now a "focal hour" would start. In a focal hour, the person on the telescope watched a single fawn, identified by the right-left color combination of the small plastic ear tags that we had put in the fawn's ears when we captured it in the week after birth. Each time the fawn had a social interaction with another pronghorn, the observer dictated the identity of the other animal, and the sequence of actions performed by each animal, to a companion, who transcribed this information in a kind of shorthand onto a data sheet. Every 5 minutes, the observer also dictated what the ongoing activity (standing, feeding, reclining, or moving) of the fawn was, as well as the identities of all of the pronghorn estimated to be within 5, 10, 20, and 50 meters. Focal observation is intense work, and after an hour, the observer was always eager to trade jobs with the transcriber. Besides being tiring, focal sampling can also seem odd. I might be doing a focal hour on a fawn that was reclined, chewing its cud, while other fawns leapt, cavorted, and butted each other; the focal procedure dictated that I record only the behavior of the focal fawn. The reason for the procedure is that it allows one to make scientifically accurate statements about the behavior of an age group or about individual differences in behavior.[2]

I wanted to provide an accurate picture of how a pronghorn fawn grows up, and I also wanted to record differences in the early experience and behavior of the fawns that I had marked. Because the plastic ear tags are very durable, the individuals whose "childhoods" I had studied would be recognizable for many years to come. I would be able to show whether certain types of early experience were associated with certain types of adult behavior. I became interested in this second question toward the end of the first field season, when Karen made her pivotal observation that fawn social status was linked to position in the birth order.

In subsequent years, just as in the first year, my assistants and I saw the same process of birth-order dominance unfold in every fawn group that we observed. We also observed adults that we had marked as fawns, to learn how the early dominance interactions that they had had in fawn social groups influenced their behavior later in life. We followed about 20 ear-tagged individuals, using the same focal procedure that we used on fawns. The results for females were starkly Calvinist: A female's dominance position when she was a fawn closely predicted her dominance status when she was a yearling, now interacting with all adult females in the population. Then, year after year, our sampling showed that an individual female's dominance position did not change. Her lifetime social status was set in early life by her position in the birth order.

This Calvinist finding brought about the watershed of my study: the learning of individual identities of every pronghorn on the Bison Range. Once I was able to recognize every individual, my view of pronghorn society expanded surprisingly. Where formerly I had seen a group of females drifting back and forth, feeding, I now saw a complicated nexus of foraging paths and subtle direction shifts. Where formerly I had seen females recline one by one after the morning feeding bout, I now saw a dramatic and often hilarious process in which each female was not allowed to recline until all females dominant over her had reclined. What's fascinating to me is that although learning individual identities in effect took off the blinders, my decision to learn identities arose from a narrow technical issue.

Here's the technical issue: I was following a small number of ear-tagged females, and I had shown that each female's dominance status was stable and was set by her fawn status. For each female, I measured status as the percentage of all her dominance interactions that she won. That, however, could be a biased measure if some females did not interact with one another. For example, if a female arranged things so that she avoided interacting with several females that were dominant over her, her status measured by the percent of her wins would give an overestimate of her true status. The only way to discover the true status of my marked females was to find their place in the overall dominance hierarchy of the entire population. And that was why I decided to learn the identities of all individuals.

How did we learn the identities of 100 unmarked animals? The first step was to find some reasonably visible traits that varied among individuals. Pronghorn have several of these, mostly on the head and neck. Some faces are light and some are dark. Horns vary in size and orientation. On the long throat, there are two white bands. The upper neck band may or may not be split in the middle. The upper neck band may or may not be connected by a white line to the lower neck band. The lower neck band may or may not be connected by another white line to the white hair on the chest. The next step was to start writing out descriptions of individual animals, noting the state of each of the variable characteristics. When we finished a description, we'd give the animal a name. I decided that the names should be mnemonic, if possible. We decided that all females should have three-letter names, and that males should have longer names that were usually complete words. Here are some examples.

Female Names

HAH   Conspicuous tufts of hair around bases of the horns: hair around horns = HAH.

BUG   This animal had a very small amount of black on the snout, confined to a spot just at the tip of the nose. Nich thought that this spot gave the appearance of a bug on the end of the nose.

TAX   In all pronghorn, a large white rump patch surrounds the stubby tail, which has varying amounts of brown on it. The brown on this animal's tail was shaped like the letter X: tail X = TAX.

WNB   This animal had very small neck bands. Nich liked the way that my 2-year-old son said the word little—"wittle": little neck bands = wittle neck bands = WNB.

NEP   In all pronghorn, the ear patch is a rectangle of white that projects down from the base of the ear. This animal had no ear patch = NEP.

DUH   This is one of the rare non-mnemonics. Karen and Jake wrote this description, at a time when Jake foresaw an impending divorce, and was spending most of his free time at the Branding Iron bar in Charlo, a tiny town 8 miles north of the Bison Range. Jake was hung over and Karen was doing all of the work needed to write the description. Finally she said, "OK, what's the name going to be?" Jake just groaned, resting his head against the side of the door frame. "DUH" said Karen.

Male Names

HANGOVER Also named in honor of Jake. This animal had a very dark face and had gray, instead of the usual white, hair around the eyes.

STICKSHIFT The way that this animal's horns curved back reminded one of my assistants of the floor-mounted stick shift on a car like an early Malibu.

NICH Named in honor of Nich. No mnemonic.

D'WHITE This animal was the son of DUH, and had a conspicuous white stripe on the center of his back.

DEXTER This animal had very even, sharply defined, symmetrical neck bands. Jake thought that he "looked like a preppie."

Although part of this work was fun, another large part was frustrating and confusing. After an animal was described and named, the next task was to reidentify the same animal on another day:

"Ok, this animal looks like DUH."

"But this animal has no connection between the lower neck band and the chest patch, and that isn't written in the description"

Could this be another animal that looks a lot like DUH? Or, when

Karen and Jake wrote the description, did they get a good look at DUH's lower neck band? Or, has DUH molted a bit more, so that a former connection now looks like a lack of a connection? We realized that something was wrong when we had accumulated 85 descriptions of females, and knew by counts that there were only 65 females on the Range. A big part of the problem, which in hindsight seems obvious, is that we started describing females in June, before they had completed the molt (loss of the old winter coat and replacement by new hair), and before they had shed their horn sheaths. Females have little spiky horns that vary in length from 1/2 inch to about 6 inches. Females shed the horn sheath, as the males do, but on a much more variable schedule. One female may shed her sheaths in June, and another female may not shed until August. The shedding may not be complete, so that the old sheath is retained for a while, stacked on top of the new sheath.

Now the frustrating process of identifying the duplicate names began. Usually the duplicates had been created because some critical aspect of the appearance, such as horn length or neck band connections, had changed. Some descriptions read: "Looks just like HAH but with small horns." In this case the two descriptions were of the same animal, before and after she shed her horn sheaths. There was also a class of females that we called "nids." Nids were of extreme average appearance, with no visible (to us) distinguishing features. They all seemed to be more or less identical, and there seemed to be about 10 of them. To learn to distinguish one nid from another, we needed to be sure when we were looking at an animal that we had described earlier. I bought a paint pellet gun. The pistol uses compressed carbon dioxide to shoot a plastic ball that contains oil-based paint. In theory, the ball breaks on impact, leaving a splatter mark on the target. In practice, the balls often bounced off animals that I shot (which was frustrating when I had spent the previous 40 minutes edging the truck close enough for a clear shot), but eventually we had temporary paint marks on all the animals that were difficult to identify. Now, when we were sure that the animal that we saw today was the one that we described yesterday, we could continue to refine the description, and could start to identify subtle idiosyncratic features, such as a tiny whorl of hair between the horns. Finally, after we had a solid description of all females, with no duplicate descriptions, we could start recognizing individuals on a regular basis. Recognition then became easier and easier, and finally we reached the point at which we could just glance at an animal and know the name, without consulting the descriptions and without really having to look at all the formal identifying features. The recognition of individual pronghorn had become like the recognition of individual human faces—an easy, automatic process. Unfortunately, this godlike ability was not transferable to new assistants.

The first data that we collected after we could quickly identify all individual females was information about dominance interactions. Now, instead of using focal sampling, we used what is called all-occurrence sampling. We still used two people, an observer and a recorder, but the procedure was even more intense than focal sampling. Now the observer called out, for every dominance interaction that occurred anywhere in the group, the name of the initiator, the name of the recipient, what the initiator did, how the recipient re-

All-occurrence sampling: Nich watches and dictates while I transcribe.

sponded, and who won. The recorder sometimes had to scribble furiously to keep up.

Two summers of all-occurrence sampling gave me the information that I wanted, and then some. First, the data showed that each female had dominance interactions with almost all other females in the population. That was possible because pronghorn groups are unstable—individuals come and go and so all possible pairs of individuals come into contact. Each of my ear-tagged females interacted with all, or nearly all, of the other females on the Range. Pronghorn groups are groups only in the sense that individuals are temporarily clumped in space, in the way that people in a train station may be said to be a group. But within a pronghorn group, the level of congeniality is much lower.

I could now make the comparison that I had wanted to make 3 years earlier: How did a female's rank when measured as her percentage of wins correspond to her rank when measured as the number of other females to which she was dominant? Answer: There was a perfect correspondence. The one-sentence question, followed by the one-sentence answer, is typical of a lot of science. A question pops into your head. You decide to address it. You work like a dog for 3 years, and then you have your short answer.

OK, I'll admit that "work like a dog" is a bit overstated. Long hours in the field allowed me to see a mother weasel enter a bluebird nest box five times and emerge, five times, with a bluebird nestling in her mouth; allowed me to see a mother badger grab her son by the scruff of the neck and run away, the young bouncing uncomfortably on the ground, because he had insisted on watching me rather than obeying his mother's commands to follow her; allowed me to hear a noise like a jet plane when a prairie falcon streaked just over my head to kill a field mouse at my feet. I like watching animals, and I can happily spend many hours watching them do almost nothing (here some assistants would probably chime in, "Lordy, let me tell you about it!").

Then too, once we knew the identity and dominance status of every female in the groups that we watched, what occurred became fascinating. When a female entered a group, she looked around, to see which others she might dominate. Having spotted a few targets, she would then quickly approach them with a bold stare or lowered head, forcing them to turn away. Meanwhile, other females in the group would have been staring at the new entrant. Several would approach her, forcing a subordinate response. Dominant females also forced subordinates to move away from feeding sites. Many of these interactions are as overt as the ones I just described, but others are very subtle, and I recognized them only after I could recognize all individuals and knew the dominance status of all.

When pronghorn forage in a group, each individual often glances up at neighboring animals, as if sizing up the potential for a confrontation. Subordinate females make slight veering turns in their foraging paths to avoid intersecting the paths of dominant females. As a result of these "vector displacements," as I called them, and the other overt bullying, subordinate females often are pushed toward the periphery of groups.

Finally, the knowledge of all identities and ranks made the observation of late-morning activity almost hilarious. In late morning, having filled the stomach and engaged in a half-dozen or so dominance interactions, each pronghorn female begins to show an intention to recline. Eventually, almost all individuals have stopped feeding and they are standing, looking at each other. Now a female reclines. This is a somewhat laborious process, in which an individual first goes down onto its "knees" (remember that the "knee" is actually the wrist), then folds the hind legs under to rest the hips on the

A female (second from left) flanked by two females who are stepping straight ahead to new feeding spots stops feeding and evaluates the competition to see if she will be able to continue on her current feeding path.

ground, then rocks onto the side, and straightens the front legs out. Promptly, a dominant female approaches with lowered head, and forces the reclined animal to stand up. The result of this interaction now ripples through the group. The individual that has just suffered this bedding displacement now looks peeved; she will probably approach and displace a standing female that is subordinate to her, and that female likely will do the same, and so on and so forth. Eventually, all individuals manage to recline, but not until there has been quite a bit of forcing up and milling around. The result of the shuffle is that subordinate females usually are reclined on the outside edge of a group.

After reading about the central aspect of pronghorn society, the snooty posturing, bullying, and violent fawn struggles for lifetime social position, you are probably wondering, as I did, about the point of all the nastiness. What do females gain or lose from their relative positions in the dominance hierarchy? My first guess was that dominant females might be more efficient at gathering food, since they would suffer few feeding or vector displacements. And food is worth the hassle.

So I designed a way to measure feeding efficiency. Once again, my assistants and I worked in teams of two. We recorded data only when we were parked quite close to the animals and all individuals in the group were actively feeding. The observer trained a zoom telescope on a focal animal and for 10 minutes dictated whether she was standing, walking, cropping (gathering plant parts into her mouth), or chewing. The other team member punched keys on a small computer (these were the days before laptops) that stored the sequence of activities and how much time was spent in each. The observer also held a thumb-click counter and counted steps by the focal animal. This work was by far the most intense and mentally tiring of the protocols that I used. Feeding pronghorn move, nibble, and clip with alacrity. To record all the transitions between walk, stand, crop, and chew required total attention and concentration. After 10 minutes, the observer was always ready to trade places with the keypuncher.

We collected over 700 of these 10-minute samples at all times of year, and after 2 years had the data to answer the question: Does dominance rank predict feeding efficiency? Answer: No. In fact there wasn't even a hint of a relationship between feeding efficiency, calculated in several different ways, and dominance rank. OK, I thought, maybe our observations of feeding behavior, even though they were as fine-grained as possible with wild animals, were too coarse. Maybe dominant animals really do get a better diet, but that can't be detected by watching them.

I realized that if I wanted to test this hypothesis, I was going to be picking up a lot of crap. I decided to assign this project to a graduate student. There is actually quite a lot of information that can be extracted from feces. Feces contain cells from the animal's gut, so with proper treatment they will yield a sample of that animal's DNA. Feces contain eggs or juvenile stages of parasites that live in or near the animal's gut, so can be used to measure an individual's parasite load. Feces contain hormones that indicate an individual's level of stress. And finally, feces contain chemicals that indicate the nutritive and caloric value of an animal's diet.

Once you obtain the feces, the analyses are straightforward. The tricky part is to watch an animal defecate, wait until it walks away, and then be able to find the spot. The graduate student that I assigned to the project became surprisingly adept at this task.

The graduate student worked with 19 females that spanned the complete range of social ranks. For an entire summer, he followed behind them, collecting his samples in little plastic bags. He looked at parasite loads and at a chemical indicator of diet quality to answer the question: Does social rank in pronghorn females predict parasite load or diet quality? Answer: No. Like my behavioral data, his chemical data showed not even a hint of a relationship.

At this point, the weird conclusion being forced upon me was that all the snooty posturing and bullying was for naught—that social dominance in female pronghorn has no practical consequence. That is weird because animals in nature usually don't do something unless there is an advantage. I decided to put the issue on hold while I continued to follow individuals throughout their lives. Eventually, I had the data to answer the question: Does dominance status predict longevity or lifetime production of young? These are the ultimate measures of whether a trait carries an advantage or not. Answer: No. In fact, remember GY, the fawn that was at the bottom of the pecking order in her year? GY became a very subordinate adult, who won only a handful of dominance interactions in her life. Nevertheless, she outlived every one of the bullies that pushed her around in 1981, and she produced more surviving young than any of them.

Now, with all of my data essentially screaming the same message at me, I had to embrace the strange conclusion that dominance in female pronghorn has no significance. Suddenly I realized that social dominance in pronghorn might be a trait like prodigious running ability. It might be a trait that was advantageous in the recent past, when cheetahs, lions, and hyenas stalked pronghorn herds, and that has been retained into the present even though it has no present utility. And once I began to admit this possibility, I was struck by one of the consequences of social rank. Remember, because of vector displacements and bedding displacements that subordinate females tend to be pushed toward the edge of a group. When dangerous predators are present, being on the outside of a group is a more vulnerable position. That has been observed in some African antelopes that are prey to modern cheetahs.[3] I began to realize that many aspects of pronghorn behavior might be relics of a different and distant past. Many aspects of their behavior might not make sense in the present.

5 *MILK POLITICS*

My mother groan'd! my father wept.
Into the dangerous world I leapt:
Helpless, naked, piping loud:
Like a fiend hid in a cloud.

Struggling in my father's hands,
Striving against my swadling bands,
Bound and weary I thought best
To sulk upon my mother's breast.

William Blake, "Infant Sorrow"

5

One of the enduring, long-term trends in the history of life is escalation. Escalation is a consequence of life that is based upon replicating molecules—the form of life that exists on earth, and the only form of life that we know. The replicating molecules are DNA. To replicate, DNA requires materials in the form of the appropriate chemical building blocks, and it requires energy. The building blocks and energy are gathered from the environment by machinery that we call organisms.[1] An organism may harvest building materials and energy that are free in the environment, but as the number of organisms increases, free materials and energy will disappear. Now, in metaphor, each organism is faced with a choice. It can increase its efficiency in gathering materials and energy, in essence competing with other organisms, or it can take a more direct approach and eat other organisms. In general, the organisms that we call plants have taken the first approach and the organisms that we call fungi and animals have taken the second. But whatever the approach, the limited amount of materials and energy engenders competition. And the view championed by Geerat Vermeij is that competition of this sort carried out over millions of years results in escalation.[2] In the arena of predators and prey, escalation means that the defenses of prey become ever more effective while the attack capacities of predators become ever more deadly. In the realm of competition and in the realm of predators and prey, escalation has meant a general increase in the metabolic rates of organisms.[3]

Animals with low metabolic rates have low rates of energy expenditure and consumption, low and variable body temperatures, and a modest capacity for activity. Organisms with high metabolic rates have high rates of energy expenditure and consumption, high and constant body temperatures, and a greatly increased capacity for sustained activity.[4] A low metabolic rate organism such as a prairie rattlesnake can crawl into its den in September, a recently eaten mouse in its stomach, and survive just fine until the following June, when it will eat its next mouse. During most of that time, however, the low rate of energy expenditure requires the snake to be cold and motionless. Compare this conservative, sluggish existence to that of a ruby-throated hummingbird, which burns energy at a furious rate and must eat about every 15 minutes to avoid death, but when it stores sufficient fat is able to fly nonstop for 500 miles across the Gulf of Mexico from Texas to Yucatan.

Among the tetrapods, the terrestrial animals with backbones, two groups, the birds and the mammals, have extremely high metabolic rates. Birds and mammals have evolved a curious condition called endothermy, in which huge amounts of energy are blown away in the form of heat. The animals use a lot of energy specifically to *make heat*, and the useful consequence is a high and very constant body temperature, which allows birds and mammals to sustain levels of activity that would be unthinkable for a reptile or an amphibian. But the cost is that birds and mammals must eat frequently to remain alive. By reptilian standards, they must eat constantly.

Two significant consequences of the escalation into endothermy are play and parental care. Parental care is not unique to birds and mammals—it is quite elaborate in carrion beetles, for instance—but what is unique in birds and mammals is the staggering amount of parental care that the young require. Endothermic young have huge energy demands because they grow rapidly (allowed by their rapid metabolic rates) and because they are hot and small and thus lose heat rapidly. In most bird species, two parents, working diligently from sunup to sundown, are required to keep up with the voracious demands of a brood.

Because of this practical constraint, most adult birds live lives that appear to be models of Christian virtue. Partners meet, court, and become attached, often for life. Partners share the work in nest building, brooding, and especially feeding the young. Therefore, ornithologists until quite recently said that most birds were "monogamous." Ornithologists now say that these birds are "socially monogamous," given that DNA paternity analyses reveal a surprising amount of fooling around (official term: extra-pair copulation, or EPC).

In mammals, monogamy, even social monogamy, is rare, because the voracious demands of the young are met entirely by the mother. The word "mammal" is almost synonymous with the word "mother" because of milk, the amazing elixir that the mammals invented. Although we often view gestation as an expensive proposition, the largest expenditure by the mother occurs after birth. The mother continues to provide all of the nourishment for the young in a drawn-out, complex sequence of events called lactation. Unfortunately, we know little about the evolutionary origin of lactation and of mammary glands, the skin glands that make milk. We do know that two other mammalian inventions, chewing and the secondary palate, likely were preconditions for the evolution of lactation. Chewing requires the specialization of some facial muscles and these allow an animal to suck efficiently (the German word for mammal is "säugetiere": literally, suck-animal). The secondary palate is familiar to you as the roof of your mouth. It is a mammalian modification of the face bones that separates the mouth from the nasal cavity, allowing one to chew (or suck) and breathe simultaneously. If you doubt that this talent is important, drink a glass of water while pinching your nose shut.

Lactation is costly, as anyone who has lived with a nursing mother knows. She can eat what amounts to a bodybuilder's diet and not gain a scrap of fat. Her idea of a satisfying midmorning snack may be a couple of burgers. And even so, the energy content of human milk is low, compared to that of the milk of many other mammals. So a female mammal, when she decides to mate, is making a huge commitment. She is committing to provide *all* of the materials and energy needed to accomplish the transformation from fertilized egg to independent offspring. In mammals, the disparity between the energy contributions of males and females is more extreme than in any other group of animals. In mammals, males provide half of the ge-

netic instructions needed to build the young; females provide everything else.

Late May is the only time when any pronghorn on the Bison
Range are the slightest bit ungainly. Females are now obviously pregnant and they waddle a little as they move about. Each female is carrying twins that together weigh 17 pounds, or about 17 percent of
her own weight. A human with comparably sized young would carry
twins that together weigh 22 pounds. But even after she has given
birth to these relatively large young, a pronghorn female has completed only about 1/3 of her total energy commitment. The remaining 2/3 of the cost will be paid in milk, and the fawns, if they survive
to weaning in late August, together will weigh 135 pounds.

Milk is complex, but the main ingredients are water, fats, proteins, sugars, and minerals. The energy content of two large glasses of
pronghorn milk is about 1,500 calories. This is about twice the energy value of cow's milk and two and a half times the energy value of
human milk. This power drink fuels the rapid growth of pronghorn
fawns and allows them to become independent of the mother in late
August, when they are 100 days old.

Milk content varies among species and it also varies within species
by the age of the young.[5] Fats, nutrient protein, and sugar content
vary, as does the level of immunoglobulins, the proteins that give
temporary immunity to the young. The milk secreted at birth and
for a few days thereafter is often called colostrum; it contains many

immunoglobulins. This secretion is gradually replaced by what is called mature milk, or mid-lactation milk. Later, there is another change to late-lactation milk, which has declining energy value. These changes in milk are accompanied by changes in behavior that represent a struggle between the mother and young for control of the milk spigot.

When a newly born, still damp pronghorn fawn first staggers to its feet, it instinctively totters beneath the mother and reaches its muzzle upward, usually poking the mother in the armpit and elsewhere before it finds the udder and closes its lips around a nipple. The mother stands very still until the fawn releases the nipple. The mother then avidly licks the fawn's anus and urinary orifice. The fawn lowers its shoulders and raises its rump, standing braced with its hind legs spread apart, and urinates and defecates into the mother's mouth. The mother swallows these offerings. An immediate advantage is that the fawn remains odorless and thus can hide effectively. A hidden advantage is that the mother can deliver appropriate immune defenses to the fawn. The mechanism works like this: If the fawn has an intestinal or urinary tract infection, the microbes that cause them will be present in the feces or urine. In the mother's intestine, specialized areas called Peyer's patches detect and bind to the microbes, and make antibodies to them. The antibodies then are transferred to the mammary gland and so into milk. The mother can, in effect, custom-design the antibodies that her fawn needs.[6] The same process occurs in other mammals, including humans, and is one of the many reasons why natural breastfeeding is preferable to formula feeding. (Most human mothers do not lick the baby's anus, but loving care provides enough transfer to make the process work).

The maternal solicitude that I described lasts for about the first 3 weeks of a fawn's life. During that time, the fawn is hiding and the mother returns to it at about 3-hour intervals. At the mother's approach, the fawn leaps up, runs to her, and dives straight for a nipple. There is no more tentative searching now—the fawn has learned where the power drink is and it wastes no time in getting to it. The mother still stands patiently while the fawn suckles to satiety, and she follows up with increasingly lengthy bouts of anus licking.

In the third week of life, a fawn spends more and more time up at these reunion periods. It nibbles vegetation. It plays. It looks around brightly at its new world, seemingly in relief after 3 hours of looking at plants at the ends of its nose. Sometimes it sees other fawns at a distance and finally the day arrives when the mother allows it to contact another mother's fawn. The two little animals approach each other tentatively, stretching their noses forward . . . and then one fawn butts the other in the neck.

Despite the initial and inevitable and never-ending aggression, fawns begin to aggregate, and so do their mothers. At this time of year, mid-June, females look a bit relieved to be back in a group. It is almost as if they have been missing all the snooty posturing and are eager to resume it. More realistically, my sense is that a solitary female feels exposed, naked, insecure, and that she endures that feeling only for the good of her hidden fawn. When fawn hiding is no longer an option (the fawn becomes too big, too visible, too impulsively active), the mother happily immerses herself back within a group. At midmorning during this time of year it is easy to spot groups of females, reclined and chewing cuds with obvious relish. Then if you look more closely and carefully, you'll see the fawn group, another knot of smaller heads about 100 to 200 yards away.

The mothers will lactate for another 10 weeks, but for fawns, the blissful period of suckling to satiety with mom at the center of the universe has ended. A fawn is only 21 days old now, but the mother begins to turn off the spigot. She becomes more likely to walk away from a hungry fawn, and now she always puts an end to suckling bouts before the fawn has finished its meal. Mothers cut off bouts with an adroit forward hop in which they swing a hind leg over the fawn's back and then bring the leg down quickly, levering the fawn's head away if the initial hop has not dislodged its hold on a nipple. This maneuver works because the fawn always suckles from the side, its neck resting against the front edge of the mother's hind leg. Mothers do not permit fawns to suckle from the rear, between the hind legs. Sometimes when the light is right I can see a thin stream of bluish milk trail from the nipple as it is wrested from the fawn's mouth. The fawn stands motionless for a few seconds, as if trying to figure out what just happened. Soon enough, however, the fawn learns what is coming when the mother begins her hop, and it scrabbles sideways to keep a few additional milliseconds of attachment. But when the mother swings her leg over the fawn's back and then snaps it down, that closes the door with finality.

A battle is now under way between mother and fawn, the fawn almost incessantly demanding, demanding, demanding, "But Mom, I'm HUNGRY," and the mother, with what looks like hard-hearted resolve, telling the fawn that it will have to look down, not up, for food. This is obviously a lesson to be learned, but the mother is also modulating affairs to suit her own interests. Sure, this fawn is important, but in less than 3 months there will be another opportunity to mate. The mother must balance the fawn's demands with her own interests in producing future fawns, and incidentally with her interest in accumulating enough fat to survive the winter. And to diminish the amount of milk that she is producing, she needs to diminish the amount of time that the fawn spends clamped to a nipple; suckling stimulates milk production. So at any encounter the mother and fawn are struggling over more than just whether the fawn may suckle right now—they are also struggling over how much milk the mother will produce in the future. Ultimately, they are struggling over how much the fawn is worth. From the fawn's point of view, the fawn (itself) is worth quite a bit, and certainly worth more than any stinking future siblings that mom is concerned about. From the mother's perspective, next year's fawns are just as good as the one that is pestering her at the moment.

In 1974, Robert Trivers worked out the evolutionary logic behind scenes such as those that I've just described. In his now-classic paper, titled "Parent-Offspring Conflict," Trivers showed that conflict between parent and offspring is to be expected, especially when the future siblings will be only half-siblings (different father), as is usually the case with pronghorn.[7]

In early August, a lactating mother is pursued closely by her 9-week-old fawn, who looks for a drink. The mother's molt is incomplete; compare her ragged coat to the sleek one of the female in the background.

I mentioned that the mother's step-over effectively ends every suckling bout because the fawn always reaches for the udder from the front of the mother's hind leg. The fawn is forced to take this approach because the mother permits none other. It would be unthinkable for a mother to allow a fawn to suckle from between her hind legs, even though this is an easier access route. Mothers require that the fawn approach from the front or the side so that, as the fawn reaches for the udder, the mother can lower her head and sniff the fawn's rump. She has been licking that rump for some time now, and she has memorized its odor. In fact, she memorized the odor in the first hour after the fawn was born. That odor is specific to the fawn and it acts like a PIN or a bar code. The mother will not permit a fawn to suckle until she has used this identification procedure to verify that the fawn is her own. During the hiding phase, when the mother and fawn are alone, the check is redundant, but after the summer groups form, the check becomes essential.

In late June, when a mother approaches a fawn group, all the young spring expectantly to their feet. They know a milk spigot when they see one, and they couldn't care less which mom it is attached to. The mother, however, views things from a different perspective. She is scrawny and hungry and looks like a beat-up mattress because of her incomplete molt. And she is that way because most of the energy and materials that she has are going into milk. Lactation is expensive. So is she likely to dispense that milk indiscriminately? No. Animals in nature are not charitable. They are programmed to act in ways that maximize their own lifetime reproduction compared to the lifetime reproduction of others. Giving milk to another female's fawn would tip the scales in the wrong direction. As the fawns throng about her, the mother uses her nose to find her own fawns, and then attempts to lead them away to suckle in peace. But other fawns are likely to follow and the mother now swings her head at them in annoyance. As her own fawn begins to suckle, she sniffs its rump again and again, making sure that this milk delivery is going to the right recipient. But occasionally she is fooled. A clever milk thief waits until suckling is under way and the mother is licking her own fawn's rump. With the mother's bar code reader temporarily engaged, the thief slips in and greedily starts to suckle. Apparently, mothers can't count how many of their four nipples are occupied. When the mother finally catches on, she whirls in fury and lunges at the thief, trying to slam it in the ribs. Fawns scatter in all directions. Pronghorn females have a way of looking peeved and they look this way most when a milk thief has just abused them.

The female and fawn groups are fairly stable throughout June, July, and most of August. If a group has attracted the solicitous attentions of a mature pronghorn buck, he has mellowed considerably after realizing that the vaginal odors emanating from females at birth did not signal a mating opportunity. In fact, the buck spends most of his time resting and eating, in preparation for his September exertions. He is for the moment an innocuous presence.

After 3 weeks of complaining, the fawns now accept that Mom has won the milk fight and they settle into a new, lower level of suckling. They still obviously relish milk, but now Mom offers only a couple of suckling opportunities per day, and she cuts off every bout quickly. A fawn is so big now that it must kneel on its wrist joints to get its head under the mother. It sucks hard and tugs at the nipple and bunts upward with enough force to lift the mother's hindquarters off the ground. The females that are lactating look conspicuously seedy compared to the females that lost their fawns to coyotes in May. A lactating female still has large hanks of hair hanging from her incomplete molt; dry females are sleek and smooth because the protein that would have gone into milk has gone into new hair. There is an economy of nature.

A fawn spends more and more time eating plants and it often forages with its head right next to the mother's. It is likely, though unproven, that this close feeding teaches the fawn something about which plants to eat and which to avoid. Fawns need to get on with these lessons because very soon they will be completely on their own in selfish pronghorn society.

6 *LITTLE SPEEDSTERS*

Gather ye Rose-buds while ye may,
   Old time is still a flying:
And this same flower that smiles to day,
   To morrow will be dying.

Robert Herrick, "To the Virgins, To Make Much of Time"

6 In summer groups, as I described earlier, fawns fight vigorously for their future status in pronghorn society. They also are still growing rapidly, at the rate of more than half a pound per day. And they are playing. It seems incongruous that they are burning energy in play at such a critical time, when energy might be put to better use in growing or fighting for status. But as July proceeds, play becomes increasingly vigorous.

By mid-July, the snipe have exited the stage of ostentatious display and the pronghorn fawns have entered. Later, in September, the pronghorn bucks will strut and fret but for now the fawns are the main attraction. And what an attraction they are! But show times are limited. Visitors to the Bison Range who have the prescience or the sleep habits that put them on the Range before 7:00 A.M. are almost guaranteed a performance. Visitors who arrive after 10:00 A.M. see only what seem to be listless animals moving slowly in the increasing heat.

Most of what pronghorn fawns do in play centers on what pronghorn do best—running. In the early morning, usually before the sun is up over the wall of the Mission Range, a fawn begins to twitch. Having just received its morning milk delivery from Mom, the fawn shakes its head as if it has a fly in its ear, then the shakes become more vigorous, and the shoulders join in. Suddenly, the fawn lowers its jaw almost to the ground and then brings it up fast. The neck is stretched upward and the movement continues into a little restrained leap, the front legs dangling straight, the head thrown back, and the rear hooves on or just clearing the ground. The fawn shakes its head again and repeats a few of these half-leaps. Now it makes a couple of fidgety-looking, nervous sideways leaps, then rocks all of its weight onto its front legs and throws the back legs around in a half circle. The adult females continue to feed busily, ignoring all this pivoting, leaping, and side-to-side jumping. The fawn dashes 10 yards or so away from the group and then dashes back, now standing in a twitching, electric, completely aroused posture. Suddenly, as if it has heard a starting gun, it sprints away from the group and begins to run really fast in a straight line. The little animal stretches its neck forward, lays its ears back, opens its mouth to gulp air, and gallops as fast as it can. The visitor who has been chuckling over the fawn's antics now is likely to gasp as the fawn begins to display the puissance and beauty of its species. The fawn's outbound course may carry it 500 to 1,000 yards away from the group before it turns in a tight loop and returns, still sprinting. It reaches the group and stands for a few seconds but then it hears another starting gun and it takes off on another, probably longer loop. As it returns to the group again, the fawn darts among the sober moms, feinting butts at them and dodging nimbly away when the peeved adults try to butt back: "Yaa Yaa Yaa—I'm quicker than you and besides that, I stole some milk from you earlier—Yaa

Yaa Yaa and watch this!" The fawn takes off on another 2,000-yard loop.

Although this snotty little 30-day-old fawn could put a thoroughbred horse to shame, it must tire eventually. But the display is not quite over. As the fawn cruises back to the group it switches to a gait called stotting, in which all four feet strike the ground simultaneously and the fawn bounds in a smooth series of energetic bounces: "Hey look at this—I've just run 2 miles in 3 minutes and I've still got something left!" Now the fawn makes a series of short dashes that are interrupted by very quick, tight right-angle turns, and now it swings in an arc around the group in an elegant trot in which the head is held proudly back and the wrist and ankle joints come up high. Finally it stops in or near the group, sides heaving, mouth open, little black tongue hanging out. I am dazzled by the displays of these little speedsters. There is deep beauty in the sight of an animal doing what it does best.

But however beautiful, pronghorn fawn play is puzzling. The fawns seem to be using energy profligately when they can least afford it. Wouldn't it be better to put all that energy into growth or into storage in the form of fat? Evolutionary biologists use the working assumption that organisms in nature spend wisely—that they partition their energy expenditures in ways that maximize survival and lifetime reproduction. I realized that before I thought more about the apparent play paradox, I needed to make some measurements of the energy budgets of fawns and of the actual amount of the total budget that they spent in play.

To make these measurements by observation, I needed to record three things. First, I needed to know the distances that fawns ran in play every day. Second, I needed to know the distance that they moved in nonplay activities every day. Finally, I needed to know the percentage of each day that fawns spent standing or moving versus reclining. With these numbers and some published information from laboratory studies of pronghorn metabolism, I would be able to come up with a fairly accurate fawn-energy spreadsheet.

To measure the distance run in play every day, I decided to focus on just a few ear-tagged fawns, which were born in the pasture where Karen and I conducted many of our observations in the first field season. The Bison Range tour road climbs a steep hillside just above this pasture, so you can park about halfway up the hill and obtain a panoramic view of the pasture. We put up a grid of flagged stakes in the pasture, and made a map, measuring the distances from stakes to other prominent landmarks. A dedicated undergraduate student then parked above the pasture every morning, and recorded the actual distances that the four focal fawns ran in play. Once per week we worked in 3-hour shifts so that someone was always watching the fawns, to record whether they were standing or reclining, from sunup to sundown.

The results? Running play takes up 2.25 percent of a fawn's total energy budget.[1] When I subtracted the energy costs of growth (relatively small) and the energy cost of metabolism (relatively huge—all that heat production) from the total budget, then play represented 20 percent of the remaining energy expenditure. In other words, once a fawn has paid the fixed costs of growth and metabolism, it burns away one-fifth of what's left on play. The amount of energy that fawns put into their morning displays is indeed significant, and I concluded that some real benefit must lie on the positive side of the balance sheet.

The benefit must have something to do with running, but what, exactly? One obvious answer is that the fawns are getting into shape —that play in effect is a series of programmed workouts. Although it is obvious and attractive, this hypothesis has a big flaw. The problem is that all of the body's getting-into-shape responses to exercise go away shortly after exercise stops. Almost everyone knows this sad fact these days. And pronghorn fawns stop playing at about 12 weeks of age, when they drift away from Mom. So if pronghorn play exists because of its getting-into-shape benefits, the benefits must be felt at the ages when the fawns are playing, or shortly thereafter. Currently, it is difficult for me to see any reason for fawns to be "training" the way that they do—to be in the kind of shape that running play must create. When they start to play as I have described, fawns already can run faster and further than a coyote.

This conclusion led me to consider two other possibilities. First, pronghorn play may have little survival value in the present, but may have had more in the recent past, when herds of pronghorn ran from lions, cheetahs, and fast hyenas. In those days, which ended just 10,000 years ago, a fawn's ability to keep up with a herd that ran from a pack of hyenas may have been critical. The other possibility is that pronghorn play has little to do with getting into shape and more to do with promoting optimal development of the brain. Just about everyone also knows now that there are sensitive periods in behavioral development, when the brain may be modified in certain ways

The half leap, drawn by the late Edson Fichter, who first studied pronghorn fawns in the Pahsimeroi River Valley of Idaho in 1956. (From R. E. Autenrieth and E. Fichter, "On the behavior and socialization of pronghorn fawns," *Wildlife Monographs,* 42 [1975]; copyright © 1975 by The Wildlife Society; reproduced with permission.)

by experience. When the sensitive period is over, the same modification is much more difficult to achieve, or impossible. Human language acquisition is a good example. Children can learn to produce the accents of a new language until about 11 years of age; thereafter, learning to produce the speech sounds of a native speaker of the new language is almost impossible. Another lesser-known example is in the development of the cerebellum, the part of the brain that controls the form and fine details of movements. If pronghorn are like other animals that have been studied, a sensitive period for the development of connections to the main output cells in the cerebellum matches the ages when the fawns are playing.[2] Play may have more to do with learning how to run than with getting into shape. Finally, I should mention that these two hypotheses aren't mutually exclusive—both could be true. But teasing these hypotheses apart will require delving into pronghorn sports medicine and brain development, invasive work that I am not keen to pursue.

I am content to watch and only partly understand the ostentatious way in which fawns burn the energy that they have drawn from their haggard mothers. Each mother would like to complete her molt and look less frumpy but, for the moment, she is giving her all to produce the next batch of milk that the fawn will drain in 20 seconds and then spend extravagantly in its next play bout.

7 COLUMNS OF DUST

In war there is no substitute for victory.

Douglas MacArthur, address
to the joint session of Congress,
19 April 1951

7

In late July and early August, when the pronghorn fawns are truly hitting their strides in play, columns of dust begin to rise across the grasslands. The bison rut has begun, and even though I am on the Range to observe another species, the bison command my attention. Most directly, they may do so by blocking the road or by denying me access to an area where I need to walk (I don't attempt to walk through a bison herd). But often, I am simply transfixed, as with the birds on Ravalli ponds, by the spectacle before me. The bison rut is a display of massive power, used in a direct and brutal way, and I find it impossible to ignore.

It is difficult to fully appreciate just how powerful bison males are, but an inkling is provided by what one can do to a human. Despite ample admonition to remain at their vehicles and not to approach wildlife, visitors often walk toward bison. Not long ago, a woman with a disposable camera approached a reclined male for a closer photo. When the bull leapt to his feet (bison agility is surprising), the would-be photographer ran for her car, and the camera proved to be disposable in more ways than one. But the male easily caught her (bison can accelerate explosively from a standing start), hooked a horn under her ribs, and casually tossed his head. She flew 30 feet through the air, and was lucky enough to escape with only lacerations and a few broken ribs (and a citation from the Range enforcement officer).

Bison males weigh about a ton (females about half that), but they are quick and agile. In the Range roundup that occurs each October, I have seen large males confined in narrow chutes make vertical leaps of 4 feet or more. The roundup corral chutes formerly were made of sturdy 3 × 12 planks, but one year a bison male ran right through the side of a chute, leaving a bison-shaped hole. Now the corral chutes are made of steel.

The bison males have massive bulk and great strength, agility, and quickness because they compete for mates in midsummer. And as is true of most of the North American ungulates, the competition is keen and swift-paced, because female estruses are synchronized— over a 3-week period all bison females destined to conceive in that year mate. The females time their estrus dates so that their calves will be born in the spring green-up, when nutritious new grass is available to support the huge energy demands of lactation. The females also achieve a level of estrus synchrony greater than that demanded by the timing of spring. Before a female comes into estrus, she shows a lot of interest in the vulva of other females, especially other females that have already mated. With this information, she times the conception of next year's calf so that it will be born just when most other calves are born.[1]

In the bison mating system, size and power count for almost everything, and the competition is stark in its simplicity. Males move

A tending bond; the male is bellowing.

through the herd, searching for females that emit a scent indicative of estrus. Prime-aged males move several miles a day in their relentless search. When a male finds a likely female, he forms what is euphemistically called a "tending bond." This means that he stays next to the female, often touching her until she is receptive and mates with him or until he is ousted by a rival. Because males can identify females that are a few days away from receptivity, the tending bond has a tiresome, obsessive appearance, the male hovering next to the female, alert to each of her tiny movements, ready at any instant to cut off a move she might make to leave, while he issues repeated threats to warn other males to stay away. Males paw large craters of dust and spray urine in them and on themselves. They roll in these smelly dust pits and the scented dust rises in a high column. Males also bellow. They stretch their necks forward and open their mouths, tongue hanging out a bit, to emit very loud, low-pitched guttural sounds that, like the columns of dust, carry a signal for over a mile. The pitch of a male's bellow is a semi-honest advertisement of his body size—the deeper the voice, the larger the male.

The bison groups tend to be larger and a bit more compact now than at other times of the year, so the activity is intense. Many columns of smelly dust rise up and drift across the herd, and deep guttural bellows sometimes combine into a rolling chorus. A male that has formed a tending bond probably will be challenged several times before his current partner agrees to mate. He periodically checks the readiness of the female by abruptly swinging his head over her rump and attempting to mount. There is nothing delicate about the mount attempt and the male actually tries to clasp the female with his front legs. An unreceptive female jumps away and over the course of a 2-day tending bond is likely to have her sides rubbed raw. A tending male also repeatedly sniffs and licks the female's vulva and inserts his nose into the falling stream when the female urinates. He then becomes temporarily frozen in the flehmen response. Many mammals have this response, in which the neck is stretched forward and the upper lip is curled upward. In bison and in most other ungulates flehmen makes the performer look quite idiotic and makes most observers laugh. Flehmen does have a purpose, however—the action moves chemicals from the mouth through a pair of tiny ducts behind the front teeth into the vomeronasal organ, a little sac in the front of the nasal cavity that performs some kinds of chemical analysis.

When challengers approach, the tending male issues a threat by bellowing. The challenger then bellows in response and a deep duet develops. In the hour before the female becomes receptive, a mature male bellows almost constantly, and probably spends quite a bit of energy just making noise. Joel Berger and Carol Cunningham actually counted bellows of bison in South Dakota, and found that in that final hour, a male may bellow 500 or more times.[2] The challenger may obtain enough information from this exchange to decide that departure is best. But if not, the males come closer. The tending male may lunge at the challenger, or he may approach, head down, tail up, at a slow walk. The two bulls then stand almost head-to-head while they nod their heads up and down. At this point one may signal submission by backing away while shaking his head from side to side, the way we do to signal "no." But if one male does not back down, the tension increases, as each male turns broadside to the other, his head held rigid and level. This side presentation in effect says, "Look at how massive I am—fighting with me would be a bad idea." Apparent size is also enhanced by the thick cape and headdress of hair each male wears, and by the luxuriant hair, called pantaloons, that covers his lower legs. The males will drop most of this hair after the rut, but for now it has an impressive display value. Also, the thick hair on the head is doubtless important for what comes next, if neither male backs down from the broadside display. The males approach head-on and collide with earth-shaking force. Sometimes, one male stands and absorbs the blow of another that crashes into him at a full gallop. The receiving male may actually slide backward 10 feet or more. Now a deadly contest is joined, in which each male tries to get his head to the side of his opponent's head so that he can hook upward with his short, brutal horns at the face, neck, or side of the opponent. A successful hook lifts the opponent's forefeet off the ground. The males drive into each other, the large muscles on their sleek rumps bunching with the effort. They roll their eyes and pant from the exertion. Dust rises up, almost blocking my view as 2 tons of muscle and lust-driven frenzy battle for the privilege of fertilizing an egg.

Eventually one male sees the better part of valor or is injured, and withdraws. The victor resumes paying close attention to the female and continues his repeated insistent preemptory mounts. The male continues to insist on a side-to-side, heads-together orientation, and he may even kick the female in the head if she disobeys. Females sometimes try to escape—these runs through the herd gather a string of males and raise a lot of dust. Eventually, the female settles for a male and she permits a complete mount. But even now she presses the issue, by running forward just as the male achieves intromission. The male adroitly follows, running on his hind legs for as far as 100 feet to keep his penis where he wants it. After a single ejaculation, the female usually loses interest, and so does the male. After a brief period of post-mating guarding, he leaves to patrol the herd in search of his next temporary obsession.

A rutting bison herd may spend several days in roughly the same area before it moves on in search of better food. In its wake, the herd leaves a heavily trampled area, studded with dust craters and circular pats of dung. When it is excreted, the dung plops to the ground in a wet tacky mass that quickly settles into a disk about the size and shape of a large deep-dish pizza. A fresh pat glistens for half an hour or so, and then acquires a duller sheen as it begins to dry. A pat that is 2 weeks old has dried into a light tan disk that is about the size, shape, and color of the hiding pronghorn fawn. These dried pats are pitted in many places where insects that spent their larval lives chewing through dung emerged to fly away.

One day, as I sat near a pronghorn group that was making its dainty way through such a bison minefield, I noticed a very fresh pat about 12 feet away. I quickly looked around to see if there was a bison nearby (yes, this can happen if a bison is temporarily hidden in a little gully). Satisfied that the former owner was not nearby, I decided to look at the pat with my 10× binoculars. When I brought the wet glistening surface into sharp focus, I found myself in another world. The dung pat was teeming with life. All over the wet top and sides of the pat, hundreds of tiny male flies were displaying. Females of the species were walking and flying from male to male and upon approach, each male elevated one wing and then the other and waltzed in a courtship dance. Their activity created a gauzy haze that flickered over the dung pat—it may have been this that subliminally caught my attention and prompted me to raise my binoculars. At less frequent intervals, yellow-eyed dung flies, about 20 times larger than the other species, patrolled on small territories and also waited for mates to appear. A pair of dung beetles industriously chewed the sides of the pat, collecting a ball that they would eventually roll way to a safe location were their progeny could feed without competition. Two very large wasps flew in loose circles around the pat and picked off flies. The last animal that I noticed was a large gray-brown wolf spider that rested motionless, centered on top of the pat. The spider made very slight anticipatory movements when little flies walked close, but while I watched for 20 minutes, did not make any captures. When I looked up at the pronghorn group still foraging delicately in the new area the bison had opened up, I saw the thousands of bison pats that covered the near and middle distance with a new appreciation for the multitude of invisible sex, life, and death struggles that came out of the larger struggles among the bison males. It was also fascinating to realize that the number of male flies struggling, mating, and dying on two bison pats was far greater than the number of bison on the Range.

Although the bison rut is transfixing in its display of brute power, there is something about it that I find well, *boring.* The male strategy seems uninspired and unimaginative: Find a likely female and stick to her with brutish glue until she consents, then find another one, and another one. . . As I will describe later, male strategies in other mating systems, especially those of pronghorn or bighorn, seem much cleverer and more inventive. The bison rut somehow reminds me of the Cold War, when two superpowers, locked into an unimaginative escalation, pointed thousands of nuclear warheads at each other. Given that force could win a contest, the only acceptable strategy to avoid defeat was to create more force. This is the sort of escalation that the bison males have become locked into in the evolutionary trajectory of the species.

There are about 350 bison on the Range. When the herd bunches together, as it tends to in the rut, the columns of dust, the mad dashes of females, and the nearly constant background of bellowing, merge into a panorama that evokes the once endless grasslands of North America. We almost shot these magnificent giants to extinction, and we plowed their grassland home from Ohio to the Rockies. On the Bison Range, if you squint a little and look across the dust-enshrouded backs into the late afternoon sun, it is possible to imagine what the Great Plains looked like at one time.

8 BACHELOR WORKOUT

I was always trying to keep up with my two older brothers while growing up. Fortunately, we were about the same size so we had each other as workout partners, which was very beneficial. We were very competitive with each other and it seemed that we were wrestling a match every time we worked out.

Cael Sanderson, undefeated four-time NCAA
wrestling champion, in an interview with
usolympicteam.com, 25 June 2002

8

Shortly after the bison festivities end, a curious burbling begins to rise from the grassland. The sounds are identifiable as meadowlark song, but they only roughly match what one hears in the spring while the pronghorn fawns are dying. It seems as if the meadowlark males are drunk. Actually, the songs are sung not by the adult males, but by their sons. The young male meadowlarks have a good idea of how their song should go, but they are not able to produce the sound yet. So they give it their best shot and listen to the sounds that come out, trying for an ever-closer match between what they hear and what they memorized shortly after they hatched, when Dad was still singing. Their songs will slowly improve over the next month before they fly south.

Meadowlark males in plastic song (the term ornithologists use to denote the period in which song is rehearsed and perfected into its final form) signal the end of the summer groups of female pronghorn and their fawns. The groups begin to drift and to break apart. The mothers were very sedentary throughout June, July, and the beginning of August. But now they dash the hopes of the male that has been accompanying them all summer by leaving abruptly. The fawns follow, still angling for the occasional drink of milk that mothers may permit. Within these groups, social relationships among the fawns now are stable and obvious. Within each sex, a linear pecking order exists, set originally by the birth order. Also, all male fawns, no matter how late they were born, are now dominant over all female fawns. The male fawns also are dominant over most of the adult females in the group, even though they weigh about half as much. The male fawns are simply more pugnacious than the adult females, who for the time being tolerate being supplanted. Soon the little hooligans will depart.

On his day of departure in late August, a male fawn simply drifts away from the group. He may veer right when the foraging group veers left, or when the group reclines, he may continue walking. Mothers do not appear to notice when their sons leave. The intense social bond that made the mother intensely vigilant, protective, and solicitous 12 weeks ago is now gone. The male wanders in a desultory way. He may join another male fawn or he may remain alone. He is now, 3 months after he emerged as a spindly-legged tottering infant that had to stretch upward to reach a nipple, a 60-pound speedster, confident in his ability to outrace any predator. He knows what to eat and his rumen, the bacterial digestive vat that provides most of his fuel, has taken on all the chemical characteristics of an adult rumen. He has little 3-inch horns and his subauricular gland, now active, is a thin black line below his ear at the back angle of his jaw. He starts to scent-mark, tentatively. He does not have all his adult teeth yet, and he will not until he is 3 years old.

After his period of wandering, the little male enters a group of other young males. He may join such a bachelor herd, as it is called, just a few days after he leaves Mom or he may not join until the following spring when the larger winter mixed-sex herds begin to break up. But whenever he joins, his social status is destined to plummet. In the bachelor herd, he and any other male fawns are now the smallest and weakest individuals, and the same asymmetries that he used to dominate every female fawn in his summer group now turn against him.

A bachelor herd contains about 5 to 20 males, aged 1/2 to 4 years. These herds occupy out-of-the way locations on the Range, usually far from older solitary males and summer female groups. "Bachelor" is an accurate description, because these males, unlike females, are in a lengthy prereproductive limbo. A female may come into estrus in her first year, when she is 5 months old, but she typically waits for a year, conceiving first when she is a 15- to 16-month-old yearling. For a male, there is no hope of reproducing when he is a fawn, and almost no hope when he is a yearling or 2-year-old. In part that is because young males are slightly smaller and weaker than mature males. But mostly it is because females in estrus look upon young males with undisguised contempt.

So until he is 3, a male's probability of mating essentially is zero. Yet a male's life expectancy is only about 8 years. Males really get squeezed—they can't mate until they're 3 and they'll be dead about 5 years later. So each male spends his limbo time doing his utmost to be ready for the window of opportunity. And that is what life in the bachelor herd is all about—preparation.

The catalog of preparatory activities that go on in a bachelor herd is meager. First and foremost, the young males eat and then they ruminate. Pronghorn are picky, dainty feeders, plucking a leaf here, a flower there, a rosehip now and then. You might look at a foraging pronghorn, its head down while it steps forward slowly, and assume that it is just mowing. Actually, mowing is a good description of how bison eat, but it is far from what pronghorn do. Get close enough and you'll see that pronghorn brush the muzzle rapidly through the grassland vegetation and nip off individual leaves, flowers, and shoots. If you are very close, you may notice that an animal runs its nose along the length of a 5-inch plant, snips off a single leaf, then moves on. But when pronghorn are foraging next to my truck, about 6 feet away, I usually get only a partial idea of what they are eating, even when I use binoculars to make the animal's head fill the field of view. They simply move too quickly while taking tiny items that are usually close to the ground and concealed by taller grass. Wildlife biologists, interested in knowing what plants are important to pronghorn, have tried to describe the diet by surveying an area just vacated by a foraging group and noting the plants that have missing parts. This task proved to be impossible because the researchers detected only a few plants that showed any signs of damage. Unlike bison, which leave a closely cropped, trampled, and dung-spattered lawn in their wake, feeding pronghorn flicker through the grassland with an almost unseen effect on the plant community.

The selective feeding allows pronghorn to have a diet that is far better than the average value of the grassland from which the diet is taken, but it is time consuming. Young males, interested in growing as rapidly as possible so that they may have an early chance to mate, spend much of their time eating. Then, having filled the rumen with tiny plant parts, they need to ruminate.

The ruminant mammals include the giraffe and okapi, the sheep and goats, the antelopes (8-pound dik dik to 1-ton eland), the cattle, buffalo, and bison, the deer (16-pound South American pudu to 1,600-pound moose), and the pronghorn. The ruminants are a successful and diverse group because of their amazing multi-chambered stomachs. The first chamber, or rumen-reticulum, is a large fermentation vat, where bacteria, protozoa, and fungi digest plants and, uniquely, digest cellulose. Cellulose is the structural sugar polymer of plant cell walls; it allows plants to stand erect. There is a lot of energy in cellulose (as evidenced by forest fires), but mammals cannot digest the long-chained molecule to unlock its energy. That's where the microbes come in. In the rumen, microbes produce an enzyme that breaks cellulose into its chemical building block, glucose. The microbes use the glucose for their own purposes, but leave energy-rich carbohydrate waste products that the ruminant absorbs into its blood through the rumen wall. Periodically, the ruminant passes a slug of the bacterial soup into the posterior chamber of its stomach, where the standard, acid-based digestion that goes on in all mammal stomachs occurs. The bacterial harvest provides the ruminant with proteins, vitamins, and nucleic acids. Ruminants also help the bacteria along with their protein assembly by recycling nitrogen. All mammals produce a nitrogenous waste product, ammonia, which is usually converted to urea, and then excreted in urine. Ruminants recycle most of the ammonia and urea into saliva, which when swallowed becomes a source of nitrogen that rumen bacteria use to make new proteins. Every ruminant animal in effect houses within itself another ecosystem, with which it lives in a balanced, intimate symbiosis.

The key to a young male's growth is the rate of digestion by bacteria and the resulting growth rate of the rumen bacterial population. This rate depends on how fast the bacteria can break down plant parts and this in turn depends on the size of the plant parts. A given mass divided into many small parts is digested much more quickly than the same mass undivided. Hold a match under a cloud of sawdust and the cloud will explode—hold the match under the same mass of wood, pressed into a log, and you'll create only a small black spot. So the ruminant obliges its rumen microbes with a double-chewing process that breaks plant parts into tiny particles. The first chewing occurs when the plant part is harvested, and is brief. Then after the rumen is filled with roughly chopped plant parts, the animal reclines and begins to chew everything again: that is rumination.

A bachelor group may spend 1 and 1/2 to 2 hours foraging on a summer morning. Then the males recline and begin to chew their cuds. The bolus of rumen material that moves up the esophagus appears as a bulge that travels rapidly up the throat. The bolus pops into the back of the mouth, the animal pauses for a second, and then begins to chew with a self-satisfied air, using the high-crowned molar teeth to shear and crush. The animal chews the bolus for about 30 seconds, then swallows, sending it back down to the rumen-reticulum accompanied by lots of nitrogen-rich saliva. Then another bolus is brought up. If pronghorn ever appear to have an air of contentment, it is when they are ruminating. We might think of an hour or more of laboriously regrinding our meals as an onerous task, but pronghorn and other ruminants I have watched appear to enjoy it. Natural selection can shape the brain to make anything that contributes directly to reproduction feel like fun.

The second activity that goes on in a bachelor group is a continuation of the dominance struggles that went on in fawn groups. Now a male fawn finds himself in a whole new league and his status drops abruptly. He may not be at the bottom of the pecking order (perhaps there is another fawn in the group that he is able to push around), but he's close to it. All yearlings dominate him, as do all older males. He begins to keep his distance from these larger bullies but still he remains in the group, because the grouping tendency is strong in pronghorn.

Grouping in pronghorn doesn't exactly make sense, because adults are never in danger. Prey animals form groups (herds of wildebeest, schools of fish, flocks of starlings) as an adaptation to predation. Join a group and your individual probability of being eaten, should you encounter a predator, is divided by the number of individuals in the group that you joined. Prey animals form groups because each individual, by joining, dilutes its own probability of death.[1] But for an adult pronghorn, the probability of death by coyote essentially is zero. So there is nothing to dilute and no real reason to join a group. Nevertheless, the young males hang around the older guys in the bachelor herd even though they suffer a steady stream of dominance losses. In part, pronghorn grouping is an evolutionary relict, like the currently overbuilt speed and endurance that evolved when the North American grassland was filled with really dangerous predators. When cheetahs, hyenas, and lions stalked Pleistocene pronghorn, being in a group would have been a very good idea. Pronghorn have not lost the urge to join in the 10,000 years since those big predators became extinct.

The males in a bachelor group assert dominance in the same ways that females do, with feeding displacements, bedding displacements, and simple displacements. And in bachelor herds, as in a fawn group, a linear pecking order, largely determined by relative age and size, exists. Pronghorn rarely resist the urge to join a group and when in a group seemingly cannot resist the urge to push each other around, even though there may be no advantage. There certainly is no advantage for females, and I have not detected a clear advantage for males either. A male might gain value from being dominant if that meant greater mating success in the future. But, as I'll describe later, mating success is determined in a heady, strenuous, sometimes violent arena that seems outside of the confines of comparatively subtle dominance assertion and submission. I think it's likely that male pronghorn push each other around for the same reason that females do—for no good reason in the present, but for probably a damn good

A bachelor herd. This photograph captures some of the heady atmosphere that is typical of these groups.

reason in the past, when being on the outside of a group meant greater exposure to some really nasty predators.

The final activity in a bachelor herd is sparring, a more or less cooperative, egalitarian, low-intensity form of fighting. Sparring is unique to bachelor herds and in the present (and possibly in the more dangerous past) provides the main advantage of bachelor herd life. Sparring is also the most noticeable aspect of bachelor herds. When bachelors are not eating or reclining and ruminating they probably are sparring. A sparring match starts when one male approaches another with horns lowered. The recipient almost always accepts this invitation and lowers his own horns. The horns touch and the males begin to push at each other and to use horn leverage to twist the sparring partner's head and neck from side to side. The sparring bout ends 30–40 seconds later when one male breaks off and turns or springs away. Either or both males may then redirect a mock attack to a nearby bush or to the ground, thrashing branches and tearing up divots. The shoving and twisting in a sparring bout can become quite vigorous, and to keep his horns engaged, a male often needs to perform some adroit maneuvers with his hind legs.

One of the most common adjustments to a sparring opponent who has gained a twisting advantage is to throw both hind legs rapidly to the side in the direction of the twist. This brings the animals back into a more neutral, head-to-head alignment. The movement is not new to young males—they performed it many times on their own when they played in a summer group. It was one of several twitchy cavorting movements that preceded the long elliptical sprints. But now the significance of the hind leg toss becomes clearer. A young male continues to use it in sparring and will perform it hundreds of times before he eventually takes up residence as a solitary eligible breeder. Several years later, when an opponent in a *real* fight has driven him partly off-balance to expose his neck to a lethal attack, the movement will be there, ready in his repertoire of actions, and will save his life. But if the movement is not performed instantly and perfectly the first time that it is really needed, the male probably will die at that moment. The same is true for a number of neck and forefoot movements that are needed to adjust to an opponent's pushing in a real fight. A real fight is very dangerous—the risk of death is about 12 percent—so training in the complex movements of offense and defense is important. Hence sparring, the constant ongoing activity, second in importance only to eating, that typifies life in a bachelor herd.

I mentioned earlier that sparring is a cooperative and egalitarian activity. What does that mean and why should males who are in training to kill each other later in life cooperate? My assistants and I devoted one summer to these questions. The herd that we studied was site-faithful and conveniently circumscribed in its summer movements. For 3 months the herd, composed of Hangover (4 years old), Stickshift (3 years old), Junior (2 years old), Violet-Orange, Larry, Yellow-Violet (yearlings), stayed at the back of Alexander Basin near the top of a long prominent feature called Antelope Ridge. With what became monotonous regularity, we'd drive to a vantage point on the road that runs along the east side of the Basin, stop, scan with binoculars for about 10 seconds until someone would say, "Yep, they're still there." Then we'd drive 2 miles out the bumpy track through the Basin to the fence that runs along its western side, then a mile along the fence to spot where we could drop off someone to watch the bachelors. Three or 4 hours later I'd drive back to pick up the observer, who had been recording identities of the participants in sparring bouts, the duration and intensity of the bouts, and the identity of the winner and loser in each bout. We recorded bout intensity on 10-point scale where 1 indicated horns just touching and 10 indicated actual fighting. The results showed that the pecking order revealed by dominance interactions among these males was identical to the order revealed by winners and losers of sparring. So there was a competitive aspect, or at least a size and strength effect, operating in sparring. But in dominance interactions, the dominant individual nearly always was the initiator of the interaction, while in sparring bouts, the eventual loser was just as likely as the eventual winner to start the interaction. Also, the average intensity level for all sparring was between 3 and 4, which indicates that the males rarely if ever made serious attempts to hurt each other. Males tended to spar most with opponents that were close to their own size, and when there was a size disparity, the older, larger male showed quite a bit of restraint, which allowed the bout to continue (too much force would cause the smaller partner to retreat). In a real fight and just before a fight, each male has his ears laid back fully, indicating an intent to kill or maim. When they spar, males hold the ears forward. Sparring is restrained and just a little bit playful, and very repetitious. The males are cooperating, for the time being, because they need sparring partners.

Bachelor herd sparring reminds me of my high school wrestling practice workouts, in which the coach made us repeat the same moves over and over and over again. All wrestling coaches will tell you the same thing—it is not enough simply to know how the move should go (my right hand goes here, my left there, I shift my weight in this way), because in a match, you will not have that much time to think. When the opportunity for a move presents itself, you must be able to perform it perfectly, and without delay. Moreover, you need to develop a body sense (as opposed to a cognitive sense) of when the opportunity is there. Repetition develops both skills.

Sparring bachelors, like the constantly burbling young meadowlark males that one starts to hear in the grassland at this time of year, have focused on some movements that they will repeat many times until the movements are perfect and locked in neural code. For the meadowlark males, the movements are those that produce song and the neural code lies in the song nuclei of the cerebrum. For the bachelors, the movements are those that a male will use when he lays his life on the line in a fight, and the neural code is mostly in the cerebellum. This kind of development, in which an action is repeated many times to improve it, explains many aspects of animal behavior, including play, plastic song in birds, sparring in the hoofed mammals, mock stalks in young carnivores, post-molt movements in some insects, and the kicking of a baby in the uterus. The sports workout is not new and it is not only an organized human activity.

9 THE TURNING YEAR

Let it be borne in mind how infinitely complex and close-fitting are the mutual relations of all organic beings to each other and and to their physical conditions of life.

Charles Darwin, *The Origin of Species*

9 Late summer arrives with signs of the turning year. Young meadowlark males burble in their full drunken glory, hawk kettles imitate columns of smoke, grasshoppers reign, prairie rattlesnakes begin the long crawl back to their dens, and the elk males start to bugle. The evening sky may expand magically. Canada geese, not yet aligned in the squadrons that will form later, straggle back and forth and begin their first plaintive honking. Pronghorn females begin to shop for a mate. At 5 A.M., Orion rears up in the southern sky. Winter, waiting in the wings, brings an occasional dusting of snow to the mountains. "I'm coming," he whispers.

On a clear still morning in early September, I noticed a thin column of black smoke rising vertically from the base of the Mission Range. The column was so compact and unwaveringly vertical that I felt compelled to look at it with my binoculars. When I brought the column into focus I saw birds, not smoke. I was looking at thousands of hawks that had found a column of rising air at the base of the mountains. Each bird had its wings set to

The approach of winter is signaled by the dusting of snow on the mountains.
The black dots in the center of the photograph are bison.

ascend in a banked, spiral path that kept it within the thermal. I
raised my binoculars to follow the column up and was amazed that it
just kept going—far higher than I could detect with my eye. Finally, I
reached the top of the column, which flared out a bit several thousand feet above the tops of the mountains. The hawks, now tiny

specks, were leaving the dissipating thermal to glide south. From that elevation, they'd be able to glide for dozens of miles before they needed to ascend again.

Columns of migrating hawks that ride these thermal elevators are called kettles. Broad-winged hawks are the most well known kettle-forming species, and in the eastern half of North America, kettles of broad-winged hawks are fairly common. But the Flathead Valley is much further west than any areas where broad-winged hawks are ever sighted, so I am not sure what species I saw.

Late summer also brings the ascendancy of grasshoppers, the dominant herbivores. Six species in three different families make up over 90 percent of all grasshoppers that may be found on the Bison Range grasslands. These are the meadow, clear-winged, migratory, red-legged, Dawson, and two-striped grasshoppers. When they hatched in the spring, the grasshoppers were tiny miniature versions of adults. Throughout the summer they have been eating and growing and molting from one adult-like stage to another. Now they've reached their final size and have become winged, sexually mature adults. Soon they will mate and the females before they die will lay eggs in the soil.

Densities of 25 grasshoppers per square meter are common, and at that density there are approximately 412 tons of grasshoppers chomping their way through vegetation on the Range, compared to only 258 tons of bison, 5 tons of pronghorn, 3 tons of bighorn sheep, 44 tons of elk, 17 tons of white-tailed and mule deer, and 8 tons of mountain goats. Thus the tonnage of grasshoppers is about 120 percent the tonnage of big herbivorous mammals. One might suspect that so many grasshoppers would decrease the quality of the range for the big mammals, but Gary Belovsky and Jennifer Slade, ecologists who were already working on the Range when I arrived in 1981, recently showed that grasshoppers accelerate the cycling of nitrogen into the soil, and thereby actually cause an increase in plant growth.[1]

In some years, when field mice are in a population boom (field mice here and in many other locales go through regular cycles of boom and crash), they may contribute another half ton to the herbivore biomass. There are also hundreds of other species of insects that are herbivorous to some extent, as well as dozens of other species of birds and mammals that eat vegetation. Thus a conservative estimate is that the Bison Range grassland supports 1,000 tons (2 million pounds) of herbivorous animal life at this time of year, as well as the annual reproductive output of all these animals.

That's a lot of energy to be produced by an assembly of plants that never grow higher than your knee. But the plants do it year after year and the grassland persists essentially unchanged, silently harvesting energy from the sun to turn air into sugar, and sugar into starch and cellulose. The herbivores eat these sources of carbohydrate and convert them back to sugar, then oxidize the sugar in the same way that plants do to release the stored solar power to do work within the body. There is a loss of energy at each chemical transformation, however, which explains why the annual production of plant biomass is at least three times the standing mass of herbivores.

The same losses explain why the 1,000 tons of herbivores support only about 2 tons of carnivores (cougars, coyotes, badgers, weasels, golden eagles, northern harriers, kestrels, merlins, prairie falcons, snakes, owls, and several dozen species of insect-eating birds). This energy and biomass pyramid, with a base represented by plants, a middle layer represented by herbivores, and a top layer represented by carnivores, is typical of almost all terrestrial ecosystems on the earth, and the numbers reflect the multiplicative costs of energy transfer as one moves up the pyramid.

The Bison Range ecosystem is stable because the numbers of herbivores are regulated. Grasshoppers are regulated by the amount of summer rain and the consequent summer plant growth, and by dozens of species of animals that eat grasshoppers. Pronghorn, deer, bighorn sheep, mountain goats, and to some extent elk are regulated by predation (mostly by coyotes) on the young. The current 5 tons of pronghorn on the Range would become over 300 tons in less than a decade if there were no infant mortality. Bison on the Range have no predators (bison mothers are belligerent guards, and even if the mother was absent, a coyote would probably be incapable of subduing a bison calf), so the population is controlled by the U.S. Fish and Wildlife Service, which administers the Range. Every year at this time the Range crew, riding horses, begins to move bison into the staging area for the annual roundup. At the roundup, individual bison are herded through a series of chutes. Many individuals are weighed, the young of the year are branded with a single digit on the hip to show the year of birth, and a certain number of adults are sold. The sale is by an earlier, sealed-bid auction. If you have a winning bid, you back your trailer up to the chute, and the crew loads a live wild bison into it. You drive away, hoping that your trailer is secure.

When I walk through the grassland at this time of year, the grasshoppers rise in small clouds in front of me. Meadowlarks, mountain bluebirds, kestrels, shrikes, and coyotes for the time being have little trouble finding a meal. Clear-winged grasshoppers make a loud clacking sound with their wings when they fly, and that sound is just enough like a rattlesnake's rattle that it startles me every time. "Stop jumping," I tell myself day after day and year after year.

But my encounters with real rattlesnakes prevent any habituation. The species on the Range is the prairie rattlesnake, *Crotalus viridis,* which is mottled green and brown and well camouflaged in the grassland. Fortunately, prairie rattlesnakes are reasonably well mannered. They don't aggressively approach, as some other species of rattlesnakes do, and when I step close to them their usual response is to crawl away quietly. But they will coil and rattle and strike if they are sufficiently provoked. So, if I step close enough to cause a rattle, believe me, I jump. I've never been bitten in 20 years of walks through the grassland, but I have occasionally stepped directly over a rattler. When a snake rattles from between my feet, I'm startled enough to prevent habituation to grasshoppers for quite some time. In late summer, I need to look carefully before I sit down. The snakes have been out all summer, feeding on field mice. Each snake has kept to a fairly restricted area, waiting day after day for its infrared detectors to go off, telling it that a warm-blooded mammal is nearby. But now summer feeding is over and the snakes need to return to their dens while the daytime air temperatures are still warm enough to allow a cold-blooded creature to move. Each snake may have several miles to travel, so at this time of the year a rattler may be encountered anywhere, not just near the ridges and outcrops where they spend the summer.

If you have been to Montana, you know that the term "big sky country" is more than a romantic metaphor—it is description. And the evenings of late summer bring on such skies in profusion. On an evening in late August, I returned from the Range at about 9:00 P.M. The sky was dark to the west, and there was a steady rumble of distant thunder. Little whirlwinds kicked up dust in the driveway. I walked into the house and found Karen sitting on the deck that faced the mountains. I sat beside her and we watched as the mountains glowed with golden light that streamed beneath the approaching thunderheads. As the sun sank below the horizon, the wind picked up and the sky above darkened to that blue of the wet sand on an evening beach after a wave recedes. Five thunderheads slid around and above us. The bottoms of the clouds looked like a cobblestone street—these were mammilary thunderheads, indicative of strong internal winds. Now the mountains darkened but the cobbled clouds picked up the golden light. Within each cloud and between clouds, bolts of silver lightning flashed against the cobalt sky. The golden light in the clouds faded to yellow, and the bottoms darkened. Now only the tops of the clouds were lit as the first stars began to appear. No rain fell.

The North American elk, or wapiti, like bison and pronghorn, once was at home on the vast grasslands that Willa Cather referred to as "the floor of the sky." When humans with plows destroyed the grasslands, the elk were forced into the Rocky Mountains, safer islands in the fast-disappearing sea of grass. Today, most hunters think of elk as a mountain species, but they are not.

Elk are members of the mammal family Cervidae, which comprises all of the varieties of deer (41 species). Male deer grow weapons called antlers on their heads. An antler is not a horn. Horns are the weapons grown by pronghorn and by all of the 140 species in the large family Bovidae. A horn is composed of a bony core surrounded by a keratinous, horny sheath. Antlers, in contrast, are entirely bone. When an antler is growing, it is surrounded by skin, often referred to as velvet; when an antler is complete, the skin dies, dries, and shreds off. At the end of the mating season every year antlers, unlike horns, are cast off. The male has freed himself from the burden of carrying a heavy hat rack on his head, but he will have to regrow all that bone next year. Thus although antlers primarily are weapons used with vicious intent to kill or maim a rival, they also by their nature have some display value. Antlers proclaim a male's ability to grow a large amount of bone in a short time (no salacious pun intended) and thus they say something about his foraging ability, his digestive efficiency,

A male
elk on
the Bison
Range,
velvet
shedding
almost
completed.

and his general metabolic health. Whether female deer make use of such information is unknown.

Male elk, their close cousins the red deer of Europe, and several other species of deer also advertise something about themselves with loud, energy-consuming songs. In red deer, the songs are called roars and they are quite guttural. In elk, the songs are called bugles and the central notes are eerily high. When a male elk bugles, he stretches his neck forward, lifts his muzzle, and opens his mouth. His sides and abdomen contact strongly and he emits a short middle-range note that quickly jumps several octaves to a very loud, pure, squealing high-pitched blast that continues for several seconds, then is followed by eight or so lower-pitched yelps. Although it seems strange to hear such a high-pitched sound coming out of a 1,200-pound animal, like the singing of a good countertenor, there is nothing about the song that seems feminine. In fact, the elk bugle carries so much energy that if I am within a quarter to half a mile from a bugling male, the hair always stands up on the back of my neck. I know of no other sound in nature that proclaims maleness with such force and purity.

The elk rut is now under way. The males are singing to advertise ownership of harems and to repel other males. Their songs ring down from the upper grassland slopes of the Bison Range. Elk songs, silvery afternoon light, and Orion rearing in the dawn sky remind me that hard frosts will be here soon and that the pronghorn rut is about to start.

The elk rut is nearly coincident with the pronghorn rut, and elk calves will be born next May and June, at the same time that pronghorn fawns are born. In other words, these two ungulate species, which differ in body size by a factor of 5, have gestation periods of equal length. That is strange. If we take into account body size, is elk gestation unusually short, or is pronghorn gestation unusually long? The answer is revealed by the sizes of newborn young compared to the sizes of the mothers. An elk mother weighs about 500 pounds and her single calf about 33 pounds. Over a 250-day gestation, the elk mother grows a calf that is 7 percent of her own mass. A pronghorn mother weighs about 100 pounds and her twin fawns together weigh about 17 pounds. Over a 250-day gestation, the pronghorn mother grows a litter that is 17 percent of her own mass. So every fall, a pronghorn mother signs on for an investment that is 241 percent of the elk mother's commitment. Elk gestation turns out to be about right for elk body mass, and pronghorn gestation unusually long. But the length of pronghorn gestation makes sense given the level of investment. In fact, the real question is how 100-pound pronghorn mothers manage to grow such a huge litter in just 8 and 1/2 months.

I don't know the answer to this question, but I suspect that the answer has something to do with pronghorn foraging and digestive efficiency. As I described in the previous chapter, pronghorn are amazingly choosy about what they eat, so my guess is that their rate of energy harvest from the grassland is greater than that of most ungulates. I don't know if this is true, but it is a testable hypothesis. Also, pronghorn are native North Americans that evolved with North American grassland plants. Herbivores and their plant prey are locked in an ever-escalating arms race of attack and counterattack. For plants, the main line of defense is chemical. Plants produce prodigious amounts and varieties of chemicals called secondary compounds that have nothing to do with plant metabolism but that do deter herbivores. Secondary compounds may smell or taste bad, and most are poisonous in some way: they inhibit digestion, block key metabolic pathways, or block the actions of hormones and neurotransmitters. The benefits and losses that we humans reap from the escalated contests between plants and their herbivores are spices, medicines, tea, and coffee (tea and coffee contain caffeine, the most widely used psychoactive drug in the world), and other poisons, such as nicotine, opium, cocaine, and THC (or tetrahydrocannabinol, in marijuana), for which humans have found some use. The main response of herbivores to plant chemical defenses is chemical counterattack—the herbivores evolve enzymes (protein chemical catalysts) to quickly inactivate the plant poisons. Pronghorn, having evolved in the North American grassland, may therefore have an especially effective battery of enzymes to deal with the chemical defenses of the plants in their diet. This is simply a guess on my part, but it is a testable hypothesis.

Although pronghorn mothers are able to make the huge annual reproductive investment that they do, I think that they are right at the limit of what is possible. There is an economy of nature. The 17 percent investment is the highest known for any ungulate; indeed, most ungulates in the 70–200-pound range have annual expenditures in the neighborhood of 5–9 percent. Year-to-year variation in the amount of rain that falls at this time of year turned out, as I discovered in 1988–89, to be a kind of natural experiment that confirmed for me that pronghorn females were indeed investing at the maximum rate possible.

The late summer of 1988 was very hot and dry. That was the summer when much of Yellowstone National Park went up in flames. On the Bison Range, one scorching day followed another, and smoke from the fires in Yellowstone and in other forests of western Montana filled the Flathead Valley and obscured the mountains. The pronghorn spent more time close to Mission Creek and drank more water than I've ever seen them imbibe. In the following spring, the pronghorn fawns were born later than usual and many were conspicuously light in weight. It seemed to me that the energy stress caused by the grassland's drying out very early had caused the growth rates of fawns inside their mothers to slow down. Spurred by this discovery, I looked at all of my data on gestation lengths and fawn weights, classifying the late summers when the fawns were conceived as average, wet (considerable rain from July through October), or dry (little or no rain from July through October). The data showed the influence of late-summer rain quite clearly. Fawns conceived in a dry summer took 5 full days of gestation longer to get to the same weights as fawns conceived in wet summers.[2] Late-summer rain is important because it delays the annual senescence of the grassland. In a dry year, the grassland is brown and dry by the beginning of August and the nutritional value of plants has declined significantly. In a wet year, patches of green, more nutritious vegetation persist into the fall. The fact that this yearly variation in late quality of the grassland had an effect convinced me that pronghorn females really were investing right at the maximum.

Then collaboration with my colleague Jack Hogg made the case even more convincingly. Jack studies bighorn sheep on the Bison Range, and his study had run longer than mine. Jack and I realized that a comparison of the effects of wet or dry summers on bighorn lamb growth during gestation would provide an interesting test of my conclusion. Bighorn have a very low rate of investment: the single lamb born in the spring is only 5 percent of the mother's weight. We predicted that if my interpretation of the climate effect on pronghorn prenatal growth rate was correct (that pronghorn are close to the max), then a species with a much lower, conservative rate of investment should not show a climate effect on growth during gestation. And that is exactly what we found. Jack's data showed that wet versus dry years made not the slightest difference in the growth of lambs during gestation.[3]

Because pronghorn females are committing to a huge energy investment every time they mate, they take the process of mate selection quite seriously, and each female expends considerable time and energy in finding a male that she deems worthy. When the elk begin to bugle and the young meadowlark males begin to burble, the pronghorn females begin to move. Male fawns have departed and the female fawns, probably still hanging around Mom, are weaned. The mothers that raised fawns now can use the protein and energy that were going into milk to put on some weight and to make their new coats. Within 2 weeks, they will be transformed from scrawny seedy hags into sleek, plump, sexually charged beings capable of provoking males into lethal contests. The speed of the transformation always amazes me, and it is an obvious reminder, year after year, of the reality of Darwin's term "the economy of nature."

The moving females drift around in small groups of three or so, but individuals come and go, so group membership changes. The groups approach older, eligible males. Each of these males has been site faithful in a very limited area all summer. Perhaps a summer group of females was with him until recently, or perhaps he was solitary on his area all summer. The females walk close to the males and stare at them, then move on. For the time being, the males are nonchalant and pay their visitors little, if any, attention. Each male is relishing his last few moments of eating and resting. Soon, there will be little time to rest and even less time to eat.

10 MAKING NEXT YEAR'S FAWNS

The rutting season of this species commences in September, the bucks run for about six weeks, and during this period fight with great courage and even a degree of ferocity . . . and while they strike with their horns they wheel and bound with prodigious activity and rapidity giving and receiving severe wounds—sometimes like fencers.

John James Audubon and John Bachman,
The Quadrupeds of North America

10

Near the end of the first week of September, the eligible males abruptly change their demeanor. Now a male uses body language and especially horn gestures to gather females that enter his area into a tightly packed group that he pushes into his hiding place. Most male areas have one or more of these special places, which offer a tactical advantage. A hiding place is usually a low area, such as a small gully, in which the female group can be concealed from distant view. The hiding place also features a spot that permits the male to stand above the female group, prepared to cut off an escape attempt, while simultaneously scanning the horizon for rivals. Before I knew about these hiding places, I was mystified in early September, when most females suddenly disappeared. How could such conspicuous animals simply drop from sight? Eventually, when I learned where most of the standard hiding places were, I was able to record the locations of most of the females in a 2-hour morning session, even though, looking across the grassland, I could see almost none.

In the language of biologists who study animal mating systems, groups of females such as those that pronghorn males pack together in a hiding place are called harems. Use of this term is not as silly as it may seem at first; the Arabic root *harama* means "to be prohibited," and that is just what each male is trying to do—prohibit other males from gaining access to the female group while prohibiting females from leaving until they mate.

But a male who has assembled and hidden a harem has accomplished little other than committing himself to performing some Sisyphean labors. He may eventually persuade some females to mate with him, but then again, he may not. If he does mate, it may be with none of the females that currently are in his harem. These uncertainties exist because females are just a bit lighter, faster, and more agile than males, and because pronghorn hands, adapted for running, are incapable of grasping. A male can only create the conditions that encourage a female to remain in his harem and can only to a limited extent thwart a female's decision to leave.

At this stage of the pronghorn rut events have a pleasing simplicity. On the National Bison Range all 60 or so females are distributed among 10 or so harems. The largest harem contains perhaps 25 or 30 females and several others may contain 2 or 3 females. Each harem male has now established a routine. The females recognize his subtle commands to go to the hiding place and they obey. The male lets them roam a bit to feed, but mostly he keeps them packed close together and out of sight. He patrols for spots where females urinated and, finding these, scrapes the ground and then urinates and defecates on top of them, obliterating the female odor. He courts discreetly within the group, testing occasionally for the presence of a female that is becoming receptive. Finding none receptive, he does not press the issue; he moves away to scan the horizon for rivals. He detects rivals at a distance of half a mile or more, and he chases these males out of sight. The females usually look up from their placid feeding to watch the chases, and while the male is gone, they also tend to start drifting apart and away from the hiding place. Having run a mile or two to evict a rival, the male hurries back to the harem and uses his subtle gestures to collect the females together and to move them back into hiding. For the moment, the harem males and their females are cooperating. The male provides the females with a tranquil existence and freedom from sexual harassment by other males. The females cooperate by obeying the male's instructions to stay close together and out of sight. The implicit contract here is tranquillity for sex, but both males and females tend to cheat. The orderly structure begins to fray.

Females are the more frequent cheaters. In fact, nearly all females violate the simple protection-for-sex contract. A female probably will stay in an initial early September harem for about a week to 10 days, reaping the benefits of the calm life that the harem male provides. The only costs to her in this arrangement are that her feeding area is circumscribed and that she may become the target of a circle chase. Circle chases tend to occur most often at this early stage of the rut. The harem male, prompted by some cue that I have yet to discover, abruptly growls and rushes at a particular female in his harem: she runs away and the male pursues, growling and matching her speed and her fast turns to stay right behind her. The turns keep the pair close to the harem. After about a minute, in which the male demonstrates his speed and agility to the watching harem and no doubt to

Early September. The females have completed their molts; notice how smooth, sleek, and well fed they all look. The male (far right) still has time to snatch a bite now and then.

the victim, he slows and allows the female to run back to where the chase started. He stands apart from the harem, sides heaving a bit, with a self-satisfied air. I have suggested that circle chases are a form of male display, but that's just a guess. In any harem, circle chases occur less frequently than once a day, so they are difficult to study. What's really going on when males do this odd chasing remains unknown.

When a female comes within about 10 days of her estrus date, she almost always leaves her initial harem. (Remember that there are no permanent social bonds among pronghorn—female groups are temporary aggregations and individuals come and go independently.) The estrus date is the day on which a female will mate. As in other mammals, estrus conjoins behavioral receptivity to copulation with the appearance of ripe eggs in the top of the oviduct. The word estrus is derived from the Greek *oistros,* meaning gadfly or goad. The Greeks thus saw the period of sexual receptivity in females as a kind of madness, in which a female acts as if she is pursued by blood-sucking flies. Among the mammals, only humans and our cousins the great apes do not have estrus. It is interesting to ponder why our lineage lost estrus, and even more interesting to imagine how estrus, if it existed, would shape human society.

A male may cheat on the early-rut cooperative contract by giving up his defense duties. When no females are close to estrus, and several males press in from several directions, a harem holder may abruptly give up and stand with apparent unconcern while the other males run into the harem, court hurriedly and indiscriminately, and chase females. The pandemonium that ensues illustrates the value of the gift that the harem male has been providing with his herding, scent-marking, and long-range vigilance. The harem females are forced to leave to avoid continuous harassment. There is only so much effort that a male will expend for promises alone.

So, by one form of cheating or by both, the orderly array of quiet harems inevitably is disrupted. A female may be in transit from one harem to another because she felt the urge to move, or because the previous harem has "blown up," as I just described. I depict females as "in transit" because the males that were in bachelor herds in the summer now act as roving gangs that fall upon single females or groups of females and harass them endlessly. If a female wants peace and safety she must quickly get to another harem where a male defends the group. Thus a female that has left a harem is on her way to another.

Why, you may be wondering, should a female leave a perfectly calm harem, where the male is dutifully fulfilling his portion of the contract, to subject herself to harassment? What's wrong with the male? I have been studying that question since 1990 and I have at least a partial answer. The answer is embodied in what females do before they mate, so let's follow WNB, a typical female, for 14 days from the moment she leaves her first harem until she mates.

On days 1 and 2, she is in a harem of 10 females held by YC, a 6-year-old male who has centered his activities on the east side of Antelope Ridge near the perimeter fence for the past 3 years. A few males have appeared on top of Antelope Ridge, about 3/4 mile away, but YC has easily run them off. Everything is calm. The harem is only loosely packed, and the females can feed whenever they choose.

On day 3 the situation is unchanged, but WNB looks antsy. She drifts away from the group several times, but YC intercepts her and pushes her back. At 10 A.M. she makes a break for it. She sprints away from the group and covers 300 yards before YC catches her. He swings around from her left to close off her escape route, but she feints right and then dodges left and continues west behind YC. YC doubles back, now growling, and rushes at her with his horns lowered. She stops behind a patch of snowberry bushes. When YC runs around the bushes, she does also, keeping the bushes between them. They run around the bushes several times, reversing direction twice, and then come to a halt, standing and staring at each other. Suddenly YC plunges into the bushes. WNB wheels and runs away, getting a lead from the few seconds that it takes YC to struggle through the brush. She heads for the top of Antelope Ridge, running hard up the steep slope. YC pursues and catches her near the top of the ridge. Both have run 3/4 mile and gained 800 feet in elevation in about 90 seconds. As YC runs in front of her, WNB dodges left, then right, then left, and gains a 50-yard lead. YC catches up and WNB stops. Both animals have their mouths open and they are panting. Down below, at the base of Antelope Ridge, YC's harem is beginning to break apart and drift away. YC lowers his horns at WNB to push her down the slope but WNB suddenly squats and urinates, then steps away. YC goes to the urine spot, sniffs it, and then jerks his head up in flehmen, his upper lip curled, his entire body for a few seconds frozen in immobility. WNB dashes out of sight during these 3 seconds. YC looks toward her, then looks down at his dispersing harem. He runs down the slope to collect his females.

WNB continues alone across Antelope Ridge and then down into the deep gully on its northern side. She runs across the gully and up to the next rise. She stops and looks right, left, and ahead. No pronghorn are in sight. She walks north, stopping to feed now and then when she encounters something tasty. After 20 minutes of slow movement to the north, she notices a harem, some small tan and white dots, about 1/2 mile to the west. She trots toward it and when she is within 300 yards, Archie, the harem male, runs out to her. He approaches her with a prancing courtship walk, his head swinging from side to side to show his black subauricular gland on both sides, while he smacks his lips and emits a soft whine that modulates to a groan. A U-shaped patch of hair that surrounds Archie's caudal gland, just in front of his tail, stands on end. WNB runs past him and into the harem. In the harem HAH and then DUH approach WNB and displace her. WNB looks around for a female that she is dominant over, but finds none. Archie comes back to the group and courts WNB again—she shakes her head "no" just the way that humans do, moves 10 feet away, and reclines.

Archie's harem is in the middle of Alexander Basin and his hiding place is not that good. It is a shallow depression that hides the group from some directions but not from others. On day 4, the harem looks calm but in the afternoon of day 5, a group of four bachelors shows up. The bachelors spread out in a ring around the harem and edge toward it. When Archie chases a male at the 12 o'clock position, a male at the 6 o'clock position runs into the harem. Archie runs back and chases him away, but now a male at 9 o'clock runs in, and the male that was at 12 o'clock has come back. Archie runs back and forth for 3 hours while the harem stands guardedly, watching events. Eventually the bachelors drift away to the north. DUH now runs away from the harem, heading south, and WNB follows. Archie makes only a feeble attempt to cut off their departure, because another female in the harem is not shaking her head "no" when Archie courts.

WNB and DUH move steadily, trotting and walking fast across the western end of Antelope Ridge. From the top, the lowering sun highlights the white rumps of YC and his harem females, 2 miles away. WNB and DUH continue south, moving on a well-worn trail that contours across the gully at the south side of Antelope Ridge and over the next ridge beyond it. At the top of this next ridge, they cross the Bison Range tour road and then continue down the back slope toward the Ravalli ponds. At the bottom of the slope, in a little ravine that is tucked in next to the perimeter fence, they encounter Fang and his harem of three females. Fang, like Archie, runs toward WNB and DUH as they approach and he courts each. They trot past him into the harem, where DUH quickly displaces two females and WNB suffers one dominance loss, but has one dominance win.

On day 6 Fang's group is calm. At 10 A.M. and 11, Fang sees a young male at the top of the ridge near the tour road, about 1 and 1/2 miles away. He emits a snort-wheeze—a loud high-pitched sneezelike sound followed by a dozen or so regularly spaced, lower-pitched chugs, and runs toward the distant male, who wheels and departs when Fang is halfway to him. The females in Fang's group spend much of their time feeding on the leaves of rose and snow-berry bushes that are in Fang's hiding place.

Early in the morning of day 7, DUH walks away from the harem, moving south. Fang moves to intercept her and, meanwhile, WNB trots north. She gets about halfway up the ridge before Fang sprints after her. Now there is an extended 4-minute bout of chasing, feint-ing, threatening, and sprinting in which Fang outperforms the best cow pony and WNB doggedly works to get away. She eventually wins, moving north. She crosses the tour road and descends into the wash at the base of Antelope Ridge. She walks up the slope of Ante-lope Ridge, passing within about half a mile of YC and his harem. YC snort-wheezes at her but she keeps moving north. YC does not pursue her. Near the top of Antelope Ridge she encounters OB, a yearling male. OB becomes tremendously excited and courts WNB, swinging his head rapidly from side to side, in a speeded-up version of what mature males do. WNB shakes her head "no" at OB and tries to run past him. But now she is up against a formidable adversary. Yearling males are only about the size of adult females; they don't yet have the shoulder or neck musculature of mature males, so they are just as fast and agile as females. Yearling males also are not suave. They court hurriedly and they don't take no for an answer. In the rut, getting caught alone by a yearling male is a female's worst night-mare. For 2 hours WNB tries to get away but OB thwarts her. The chasing back and forth goes on out of sight of YC's harem, but slowly WNB manages to edge toward the north side of the ridge and finally into view of Kareem, who holds a harem at the base of Antelope

Ridge, on the north side, about three-quarters of a mile north and out of sight of YC. When Kareem sees the pair, he snort-wheezes and then runs toward them. OB makes a last desperate attempt to drive WNB south, and then runs away as Kareem sprints up the ridge. As Kareem runs past WNB, intent on goring OB in the rump, WNB runs down and joins the harem, and is quickly displaced by three females. Kareem returns 15 minutes later and courts WNB. She turns away and reclines.

Later in the afternoon of day 7, a group of five bachelors moves in on Kareem's harem. Kareem struggles mightily for 3 hours to repel them but the situation becomes increasingly tense as a ring of males tightens closer and closer. Finally, when one of the bachelors runs into the harem, all the females abruptly leave. Three of them head west toward Archie, two move north, and WNB moves south, staying close to the road and the perimeter fence. She's not going to risk running into OB again. Half an hour later she's back in YC's harem.

On the morning of day 8, YC walks into his harem and swings his head back and forth, courting no female in particular, but trying to identify females that may be close to estrus. One of his head wags occurs in front of WNB. She turns away. Later in the morning, WNB leaves YC, this time without a lot of resistance, and moves north. Twenty minutes later she reaches Kareem, who tries to herd her into his hiding place, where 9 other females are standing. WNB resists and Kareem is unable to prevail. The two are running back and forth on a flat area just above Kareem's hiding place, and they are in full view from many directions. After 10 minutes, Kareem lowers his horns and runs at WNB to drive her away. She trots north and

Kareem returns to his harem. WNB covers the next mile in an easy rocking canter, and within 4 minutes enters GB's harem. GB is a 7-year-old male, near the end of his life. He has had a successful reproductive career, and females still find him attractive. He is a conspicuously squat male, with shorter than average legs, a massive body, short horns, and large, grizzled subauricular glands. GB's slight stubbiness makes him particularly agile but does not compromise his speed. GB watches WNB enter his harem but does not immediately respond. Ten minutes later, GB casually approaches WNB and courts. She shakes her head "no." The afternoon is calm in GB's group. Another female, TAX, is starting to allow GB to court beyond an initial approach. She stands as GB approaches her from the front, head wagging, lip smacking, and groaning, and continues to stand as GB sidles behind her. When GB rears up, she steps away. No other males approach the harem.

On the morning of day 9, WNB runs away from the group while GB is courting TAX. GB pursues, but as he does, TAX begins to walk north. GB gives up his attempt to restrain WNB, and runs after TAX. WNB continues southwest, into the middle of Alexander Basin. She encounters a group of three bachelors. They run at her and she stands, looking frightened. When Junior courts her, she runs south. The males follow and now WNB stretches her neck forward and runs at top speed. The bachelors pursue and stay close behind. WNB runs into the wide gully at the bottom of Antelope Ridge and stops next to a large patch of brush. The bachelors come over the lip of the gully 10 seconds later and for a moment cannot find her. But the ruse gives WNB only a 30-second respite. The bachelors find her again and the chase resumes. WNB runs up Antelope Ridge, angling over the eastern side to cross more easily, and runs directly to YC's harem, closely pursued by the bachelors. YC runs at the bachelors, but they fan out. When YC chases one, the other two approach the harem. YC comes back to the harem and herds the females into a very tight group. The bachelors press in and YC is able to make only short rushes at individual males. Within 10 minutes, the harem blows up. A male runs into the female group, lowers his horns, and chases two females. The group scatters and then another male runs in, and then the third. The males are trying to force little groups of females to leave and they are courting hurriedly, and they are threatening each other. YC cannot control the situation. WNB runs up the gully on the south side of Antelope Ridge and then directly up and over the high point of the ridge. Other females scatter in several directions, and 15 minutes later YC is alone, his harem gone because of the males that WNB enticed in. WNB follows a path close to that she took a few days ago, and in 30 minutes has arrived back in Archie's harem in the middle of Alexander Basin.

On days 10, 11, and 12, WNB remains in Archie's harem. Archie easily runs off the occasional male who approaches. LEN, a female in Archie's harem, enters estrus and on day 12 Archie mates with her.

On day 13, Archie courts WNB and she does not immediately reject him. Archie gets behind her and rears up. His penis slides out of its sheath. The pronghorn penis is pink, about 5 inches long, and shaped like an ice pick. Archie is balancing on his hind legs, his forelegs hanging on either side of WNB's flank, not touching her. The tip of his penis bumps into her tail and she jumps forward, and then walks away. This sequence is repeated 47 times during the rest of the day. Sometimes WNB stands when Archie rears up, sometimes she only stands until he starts to rear up. Twice, Archie appears to become annoyed with WNB after half a dozen or so failed mounts—he growls and lowers his horns at her.

On day 14, Archie is courting WNB almost continuously. Now, when he rears up, WNB stands still for about 3 seconds before she steps away. With each courtship approach, WNB seems to stand just a little longer. But she still does not allow Archie to rest his chest on her rump and she still holds her tail firmly clamped over her vulva. Nevertheless, Archie's chances are looking pretty good. WNB has been in his harem for 4 days and she is not trying to escape. She is slowly becoming more receptive. If Archie can just keep everything as it is, he should be able to mate with WNB within 24 hours.

But although Archie's strategy is to keep everything as it is, WNB's strategy is different. She now has started to release a sex-attractant odor from her vagina. The odor is volatile and it drifts easily on the mild west wind. The odor reaches Kareem, 1 and 1/2 miles to the east. For the moment, Kareem is alone; he lost his harem yesterday. Kareem follows the odor west and within half an hour he is standing on a low rise, looking at Archie's harem 200 yards below. Kareem has made a careful approach so that only his head peeks above the rise. He studies the situation for several minutes and watches while Archie mounts and WNB steps forward.

Kareem makes up his mind. He steps forward into full view and begins to walk toward Archie's harem. Archie sees him almost immediately and snort-wheezes. Kareem keeps walking. Archie does not run at Kareem; he walks, holding his ears forward and his head up. The two males draw within 20 feet of each other. All the females in the harem stop feeding and watch. Archie turns broadside to Kareem, stretches his neck forward, and flattens his ears back. Kareem does the same and the two males walk parallel to each other for several yards. The males are moving in a slow, deliberate way and the atmosphere is charged. Archie turns toward Kareem, his horns lowered, his nose almost touching the ground. Kareem swings his horns toward Archie and the horns touch with an audible clack. Suddenly, the two males are moving their heads very fast, up and down and side to side, fencing for an advantage. Archie slips his horns past Kareem's prongs. He lunges forward, his right horn tip gouging Kareem's neck. Kareem kicks his hind legs around to the side and regains a horn-to-horn alignment. Kareem drives forward hard and pushes Archie partly off balance. Archie recovers and swings his back legs around so that he is on a little hill. He uses the gravity advantage to drive into Kareem and push him back. The action is fast and brutal and it makes sparring seem like child's play. Kareem thrusts upward, disengages for an instant, and then drives his horns under Archie's chin. The horns are forced up and around Archie's neck, and now Kareem jerks his head up and down with great force, his horn tips gouging into the back of Archie's neck, the prongs cutting up into Archie's throat. Caught in this position, Archie can do little other than try to retain his balance and wait for an opportunity to break free. After 2 minutes of continuous pummeling, Archie manages to twist his neck free when Kareem's right horn tip slips down. The two males spring apart. Archie lunges at Kareem, pushing him back, and

Kareem dodges to avoid being gored by Archie. Excellent light allowed me to stop the action of two males running at 50 miles per hour.

back again. Kareem seems to be tiring a bit, or perhaps he is hurt. Suddenly, Kareem breaks off and runs away. Archie runs after him for a short distance, and then returns to his harem. Kareem walks east, back to his spot. He is limping, and does not rest any weight on his right front foot. WNB and the other females watched the fight. As

Archie returns to the group, he is panting, and he has several large hanks of hair hanging from the back of his neck.

In about 5 minutes, Archie has regained his composure. He courts WNB and as usual she steps forward just after he rears up. He stays behind her and mounts again and again, sometimes walking for several yards on his hind legs to follow WNB. Now as he mounts, WNB's tail is starting to flick to the side. One hour later, it seems that mating is imminent. With each mount WNB takes just a tiny step forward, but the tiny step is sufficient to prevent Archie's penis from reaching its target. In moments like this one, I can imagine a pronghorn male wishing for grasping hands.

Junior, a 3-year-old bachelor, comes into view. Like Kareem, Junior has followed his nose to the harem. Archie makes a short rush at Junior, but then hurries back to WNB. Junior edges closer, and again Archie runs for only 30 yards or so to push him back. It is clear that Junior will not go away unless decisively chased away, but Archie seems unwilling to leave WNB for that amount of time. Archie tries another mount and then another, and Junior approaches boldly to within 10 yards. Archie, standing directly behind WNB, ignores him for the time being. Archie rears up and his penis slides out of its sheath for the 150th time today, but WNB suddenly runs directly toward Junior. Archie pursues and as he makes a short dash at Junior, WNB veers away and gets a substantial lead. She streaks across Alexander Basin, Archie and Junior in hot pursuit. In less than 2 minutes, she enters GB's harem of 8 females.

GB courts WNB, who takes only a couple of minuscule steps forward. Now Junior and Archie appear. GB calmly walks toward them. The two male intruders spread apart a bit. GB looks at each, pauses for a moment, then lowers his head and walks toward Junior. Junior turns quickly and runs away about 100 yards. GB then turns and walks deliberately toward Archie, with the same head-down posture. Archie stands his ground and GB continues to advance without let-up. When GB is 6 feet away, Archie lowers his head and in the next instant, the two males are fighting. Within 10 seconds Archie is driven backward and almost knocked off his feet. He breaks off and runs away, GB pursuing for about 80 yards. GB now looks toward Junior, who is walking toward the harem. GB canters toward Junior, who turns and runs another 150 yards away. GB looks at Archie, now standing 300 yards from the harem. GB walks deliberately toward Archie, who now turns and canters west toward his harem. GB looks in Junior's direction—Junior has moved into a gully to the north and is out of sight. GB lopes back to the harem and stops about 10 yards in front of WNB, who has been watching the proceedings. GB stands stock still for about 30 seconds, and then begins a courtship approach. As he reaches WNB he holds his head up high and WNB, mouth open a bit, sniffs at GB's large grizzled subauricular gland. GB works behind WNB and rears up. WNB's tail flips up and her head leans back. GB rests his chest on WNB's rump. She is standing rock-solid, head back to brace against his weight. GB is making little probing movements with his penis. The tip finds the entrance to WNB's vagina and the penis quickly slides completely in. GB ejaculates instantly, throwing his hips into WNB, his head snapping back so that his nose points to the sky.

WNB steps forward. GB lowers his horns at her and she walks off and reclines. Over the next hour, GB continues to court WNB, but does not get another intromission. In fact, with each courtship approach, WNB's response shows that she is less receptive. Two hours after the mating, WNB casually walks away from the harem and GB does nothing to stop her. As she walks through Alexander Basin, several other males see her, but they do not approach. The volatile scent that WNB had been releasing now is gone, or perhaps it has been mixed with a scent from GB, and now carries a new message.

In any year, most females act like WNB.[1] They move independently and they move from harem to harem at an increasing rate as they approach estrus. Their coming and going creates many tactical problems for males, and males who cannot deal with the problems lose their harems. A female close to estrus will not remain in the company of a male who fails to defend a large zone around her. Eventually, each female mates once. All estruses occur within about 12 days in the latter half of September, so during that time many females are moving back and forth and the popular harem holders face many challenges. Successful males, those that manage to mate with several females in one year, are the ones who maintain large harems throughout most of the rut. This means that successful males are healthy, vigorous males, because the sustained, day-after-day running, patrolling, herding, courting, scent-marking, snort-wheezing, threatening, and fighting is taxing.

The actions of females, more than the actions of other males, force males into these exertions. In a very real way, each female carefully selects the sire of her next fawns. As a result of female choosiness,

This female has made her choice. Her head is held back, to brace against the male.

one half of all males that survive to weaning will never mate in their lifetimes. Also, the stringent assay that females use has clearly put a brake on the evolution of male size. A huge male might win all his fights, but he would not be fast enough or have the endurance to chase off rivals, and he would not have a hope of controlling a harem

of flighty agile females. Ancient predators obviously created the selection pressures that shaped the running abilities of pronghorn, but
female pronghorn continued to sculpt those abilities, because they
were selected to choose mates that gave them speedy offspring. As a
result, males remained lightweight and delicately built. For males, a
serious consequence of this is that fights are often lethal. In pronghorn, the separate and sometimes conflicting interests of the sexes
play out in a dynamic, almost theatrical way. Can you see why I find
the bison rut boring?

11 AFTER THE EQUINOX

Wilbur was frantic. Charlotte's babies were disappearing at a great rate.

E. B. White, *Charlotte's Web*

11 A morning arrives after the equinox when suddenly there is silence. The young meadowlark males have flown south, a hard frost overnight has killed all the crickets, and pronghorn males no longer snort-wheeze. A few ballooning spiderlings float by soundlessly, illuminated by the silver fall light. The silence is startling at first—it makes me aware each year of the rich background of biological sounds to which I have become habituated. Now the last pronghorn female has made her choice and the air is free of male-maddening scents. Almost overnight the pronghorn males become chummy. Two males that held harems a week ago and would at that time have tried to rip each other's throats out now feed side by side. Unlike humans and several other primates, pronghorn never appear to carry a grudge. They are creatures of the moment and of immediate expediency.

For the past two weeks the males have had little time to eat. (During the peak 2 weeks of the rut, a harem-holding male is able to spend only 20 percent of his time feeding, barely enough to keep his rumen ecosystem from collapsing; during the same time, the females, guarded from harassment by the male, spend 40 percent of their time

Facing page: After the rut, a male looks for something nutritious to eat in the dry grassland. Males at this time of year are at the low point of the fat cycle. Two points of the male's pelvis are visible beneath the skin, and the rump is slightly concave. Compare the plump, rounded rump of the female in the background.

feeding.) Now they need to recoup their losses as fast as possible and they feed with an intensity that I see at no other time of year. A male that held a large harem throughout the rut has burned so much energy and has eaten so little that his fat stores are nearly depleted. To survive the winter he needs to replenish this fat, and that means keeping his rumen factory humming at peak capacity. Now the supreme importance of teeth becomes apparent. There are few succulent plants left, so each male must grind his way through a lot of fibrous vegetation.

The groups of busily feeding males grow larger and then the females come out of hiding. Many females after estrus simply disappear, moving to remote safe locations where they won't be harassed. Toward the end of the pronghorn rut I'm struck by how female-free the Range seems. Since the 30 square miles of the Range are enclosed by a perimeter fence, I know that 60 or so highly visible females are out there somewhere. But they're not in the grassland and not in harem hiding places. They are tucked away in little pockets at higher elevations, close to and sometimes even in the timber.

My searches to find post-estrus females stemmed from idle curiosity coupled with wounded pride. I thought that I knew the topography of the Range pretty well, but many females were simply disappearing. I wanted to know where they were going and I obviously needed to fill in some holes in my mental map. My searches took me to the south side of the Range, where several little canyons with names such as Telephone Draw, Twin Canyon, Trisky Creek, Elk Creek, Firehole Canyon, and Kicking Deer Draw rise from the Jocko River to the timber near High Point. This is the most secluded part of the Range and also one of the loveliest. Grassland is punctuated by magnificent solitary ponderosa pines, and higher up, the pines grow closer and closer together, forming a forest. The pines follow the canyon bottoms down to the Jocko. On golden afternoons near the river and usually close to a canyon and its little grove of pines, I'm likely to encounter a group of Lewis's woodpeckers. About the size of small crows, they are curiously silent and shy, emanating an aura of mystery. Captain Meriwether Lewis, for whom they are named, wrote of them:

> Arround the base of the beak including the eye and a small part of the throat is of a fine crimson red. the neck and as low as the croop in front is of an iron grey. the belly and breast is a curious mixture of white and blood reed which has much the appearance of having been artifically painted or stained of that colour. the red reather predominates. the top of the head back, sides, upper surface of the wings and tail are black, with a gossey tint of green in a certain exposure to the light. the under side of the wings and tail are of a sooty black . . . the eye is moderately large, purple black and iris of a dark yellowish brown.[1]

Trisky Creek, on the south side of the Bison Range, in early October.

These birds never let me look at them for long. They're just passing through and they probably have stopped near the Jocko to feed on the insects that fly around the water. I always see Lewis's woodpeckers in a small group, and the individuals always stay close together, maintaining their cohesion without sound, or at least without sound that I can hear.

Not far from where I run into Lewis's woodpeckers, I'm likely to find a group of female bighorn sheep and their young of the year. In this protected population, where humans are never a threat, the sheep are amazingly tame. I always take pains to pass groups at a respectful distance, but the ewes' reaction suggests that they'd be unconcerned if I walked among them. I'm always a bit startled when I spot a group of sheep. I'm looking for pronghorn and I have that four-chopsticks-in-a-bratwurst search image in mind, and then I see an ungulate that's about the right size, but is much stockier with almost stubby legs. It takes me a moment to realize what I'm seeing. (It can be really bizarre if, when I've been alone for several hours, I spot a human standing or walking in the distance. For an instant, my reaction is, "What the hell is that"? A large mammal that walks on its hind legs seems very strange.) The sheep mothers had their lambs here on the south side of the Bison Range in the spring, at the same time that the pronghorn fawns were born. But although the pronghorn fawns are now weaned and have left their mothers forever, the sheep lambs are still in their maternal groups. In fact, the female lambs will stay in their mother's group for many years. The male lambs will eventually leave and then will wait even longer than a pronghorn male for the opportunity to mate.

Higher up in the canyons, usually amidst the scattered ponderosa pines, I hear and then see the mature bighorn sheep males. I hear them first because they are fighting for dominance by ramming each other head on. The collision produces a loud hollow TOK that echoes around the nearby hills. When I see the two males they are standing a few feet apart, looking at each other. One male is a rich chocolate brown and the other is a much lighter tan. Both have full-curl horns that form massive outwardly spiraling arcs of more than 360 degrees. Each male has a broken or splintered horn tip, but the tips are not that important. The males stand facing each other for several minutes. One ram then backs up and the other follows suit. Both animals rear onto their hind legs and then lunge forward, head slanting down at the opponent. The rams push hard from their hind legs and the thick fronts of their horns collide with brutal force. When I am close to the combatants, the sound of a clash is almost like a rifle shot. Someone should outfit ram horns with data-logging accelerometers. I'd bet that the recorded g forces would be far greater than those that turn on a car's airbag. For a few seconds after the collision, the two males look dazed. Probably they *are* dazed. Then the males stand and look at each other again. These clash fights are curiously formal. They have none of the "I'm going to kill you in any way that I can" nature of fights between pronghorn males. But it is clear that in the evolutionary past, bighorn males experienced sustained sexual selection to win these fights, because their horns are astoundingly massive.

The horns of bighorn, pronghorn, goats, cattle, buffalo, bison, and antelopes are double structures. An inner horn core of bone grows from the skull bones. The core is covered by and firmly attached to the outer horn sheath, made of the fibrous proteins that create hair, hooves, nails, and claws. In bighorn sheep, the sheath is thick, dense, and quite heavy. The skull bones under the horn core have a unique latticework of air cells that act as shock absorbers. KA-POW! The males rear up and collide again. They will do this many times until one of them gives up and turns away. The other ram will pursue, kicking at the loser to reinforce the victory.

As the males turn away from me, I see that they could just as aptly be called bigball sheep. They have huge testes that look every bit as exaggerated and out of proportion as their massive horns. This seems incongruous, if the males are settling dominance with formal fights, and dominance will settle mating contests. And indeed, until the recent age of reasonably easy genetic paternity assignment, the size of bighorn testes *was* incongruous, because it seemed that the large, dominant rams dominated access to ewes in estrus.

Then Jack Hogg started his study of bighorn on the Bison Range. Early on, Jack showed that the bighorn rut was not as simple as once was thought.[2] Ewes do move to areas where dominant rams control access to them, but now a critical feature of female biology comes into play. Unlike pronghorn females, bighorn females accept dozens of matings. A large dominant ram mates repeatedly with a ewe in estrus, but if he does not mate with her every time that she accepts a male, he may not fertilize her egg. Jack showed that bighorn males that were not in the tending ram elite used two other strategies to compete for matings. Blocking rams intercept solitary ewes far from the center of rut activities and prohibit further forward movement. Some ewes, blocked in this way, eventually spend their estrus with the blocker. Jack called the other strategy coursing. Coursing rams hang around tending rams and their ewes and wait for an opportune moment to dash in and separate the pair. Faced with a sudden melee, the ewe runs and the coursers pursue, leaping onto her back and sometimes managing to ejaculate on the run. Because ewes mate many times and because they may mate with more than one partner, a male's reproductive success is influenced by how well his sperm compete in the pool of sperm inside the female. One simple and effective way to compete is with numbers: The more sperm a male deposits in a female, the greater the chance that one of his sperm will fertilize the egg. That's why bighorn sheep have gigantic testes.

Jack used genetic analysis to confirm that blocking and coursing were successful at least occasionally. The blocking rams had a low percentage of paternity, as one would expect—ewes often manage to escape a blocker. But the success of the coursing rams, those mate-on-the-run guys, was an astonishing 44 percent![3]

This high success rate means first that coursers somehow "know" the best time to intervene and second that ewes must be encouraging the success of flying copulations. As is true for pronghorn, the actions and choices of females have a profound effect on the biology of males. Pronghorn females choose mates by conducting an assay for male stamina and vigor, traits that were useful to offspring millennia ago when the North American grassland was full of dangerous predators. Bighorn females in part choose mates for their horns and clashing ability, but by favoring the success of coursers, they also conduct an assay for male agility, a trait useful to offspring in a species that escapes predators by moving to inaccessible cliffs and tiny ledges.

Back in the grassland, most pronghorn females have come out of hiding and have formed larger groups. The female groups merge with male groups and now I may count 50 or 60 individuals together, all feeding calmly, the concerns of summer and the rut now past. They look conspicuously relaxed and the change in their demeanor demonstrates how keyed up they were during September. Now hard frosts come almost every night, and in the morning fog rises from the Flathead River and spreads up the slopes of the Range. Frost coats each grass stem and grows in exaggerated spikes from the barbed wire on fences. The sun breaks through at midmorning and for now, at least, the afternoons are mild.

As the fogs lifts and I am able to see all the group members clearly, I notice that some of the males are missing horns. The lethal meat hooks that they carried throughout summer and rut are gone, replaced by short simple spikes. The males are discarding their used, often broken and blunted horn sheaths and are growing new ones. Pronghorn are the only horned mammals that do this. The new horn sheath starts to grow beneath the old one and finally, at this time of year, pushes it off. Now the new sheath is just a simple blade, conforming to the shape of the horn core inside it. During the winter, the new sheath will grow and acquire its prong and its deadly hooked tip. But for the time being, the males blend in well with the females.

Every pronghorn female, except for the female fawns, is pregnant. Each yearling female, now 17 months old, was panicked by her first estrus, but hormones gave her the courage to stand when she had to. Older females that weaned huge twins are pregnant. Fourteen-year-old females, who probably won't make it through the winter, are pregnant. Males show their mettle in the amazing sustained effort required by the rut; females show it by their unrelenting energy commitment, year after year. Especially at this season, my heart goes out to the pronghorn. They are exposed to every kind of weather that the Range has to dish up, they are about to enter a season of slow starvation, and they will struggle to reproduce until the day they die.

Inside each pregnant female, the struggle has begun. Although the female will bear twins next spring, she produced 4 to 7 eggs at estrus and all were fertilized. Each embryo is now about the size and shape of a short piece of thread and it is floating freely in the uterus. As the embryos come into contact they become tangled together; the twisting and knotting causes some to die. If four embryos survive the tangling and knotting, all implant. The pronghorn uterus, like the uterus of many mammals, but unlike the human uterus, is divided into two lengthy sides, called horns, which meet just above the cervix. Now each horn has an embryo implanted close to the cervix and another implanted near the top of the horn where the fallopian tube enters. But in both horns, the blood supply is better at the cervix end than at the fallopian tube end. The embryo with the better blood supply grows faster and it pushes a rotting tip of its membranes toward the other embryo. The tip pierces and kills the other embryo. Now there are two embryos, one in each uterine horn.[4] One advantage of this grisly mechanism is that a balanced uterus, with one fawn on each side, is guaranteed.[5] This kind of balance is important if a pregnant female often has to run from dangerous predators. Of course, expectant pronghorn mothers no longer have to contend with cheetahs, or lions, or huge hyenas. The disposable embryos, like amazing running

ability, dominance striving, the urge to be in groups, the intense female scrutiny of mate vigor, are relics of that more dangerous past.

The pronghorn groups grow steadily larger and the days grow
shorter. Now the ground is hard and the morning fog may not dissi-
pate until midafternoon. Snow is heavy in the Mission Range and it
begins to settle in the upper forests of the Bison Range. The rattlesnakes are balled up in their dens. The bighorn sheep are starting
their rut. Short-eared owls hover low over the drying grassland to
snatch voles. Soon the clouds will descend and for weeks at a time
the mountains will disappear. The valley is filled with a soft white
murk, and all of the pronghorn, when I can see them, have formed a
single large group. Snow falls on the grassland, completing the blan-
ket of silence. Winter, the season of marking time, slow starvation,
odd tranquillity, and death, has arrived.

12 **AFTER THE SOLSTICE**

I prefer winter and fall, when you feel the bone structure in the land-scape—the loneliness of it—the dead feeling of winter. Something waits beneath it—the whole story doesn't show.

Andrew Wyeth, interview
with Richard Meryman in *Life,*
14 May 1965

12 Winter is the only season when I can clearly see what pronghorn eat. Snow covers the grassland and only the tallest dried grass stems are visible. The animals plow slowly through the snow and clip these stems. They are eating straw. Snowberry and wild rose bushes may extend above the snow, but these plants grow in lower swales and wash bottoms where the snow is deeper. The pronghorn stay out of deep snow when they can, to keep their running advantage over coyotes, so their diet has a grim monotony.

At the solstice, each adult is fatter than it will be at any other time of year. The steady, intense feeding after the rut extracted the most energy possible from the drying grassland, and now the pronghorn enter a period of starvation and begin a rapid slide toward death. If an individual has stored enough fat, it can use these energy reserves to slow the slide and survive until spring. If it has not stored that much fat, survival depends on luck—a mild winter will allow survival, and a harsh winter will bring death. In the economy of nature, fat is good. As she starts her slide, each female carries two embryos, one in each horn of the uterus. Each embryo weighs 1/7 ounce and is 2 inches long.

To save energy, the pronghorn have grown their hair out. The hair is now at its maximum length and it is thick—almost bushy. The individual hairs are stiff, brittle, and waxy. Each is a waterproof hollow tube that is divided internally into many tiny air pockets. The key to good insulation is dead air space, and pronghorn hair provides as much as a thick down coat. Thus pronghorn in winter don't really feel the cold until temperatures drop below zero. On many winter days, I am impressed by how comfortable and relaxed individuals look, reclined in a little cavity in the snow, quietly chewing cuds. The main challenge of winter, then, is not the increased energy cost of keeping warm in the cold or the increased energy cost of moving through snow, but the decreased energy intake that comes about because all the plants have stopped growing. The annuals have died and the perennials have sent all of their energy reserves out of reach into their roots.

When they are not reclined or ruminating, the pronghorn are up and eating. These are the two activities of winter, and they follow each other with what one researcher in Alberta called monotonous regularity. The foraging periods are very long now, because the grass stems are scattered, because movement slows in the snow, and because, as the snow deepens, the pronghorn are forced to paw craters to get down to something edible.

The winter groups are large and on the Bison Range the entire population of 120 or so may be in a single group. In other populations, researchers have counted 1,000 or more animals in winter groups.[1] These big winter groups were intriguing to me, because they seemed to violate some commonly held beliefs about grouping. The common view is that grouping represents a trade-off of benefit and cost. The benefit is a reduction of the risk of death by a predator. The cost is an increase in the level of competition for food. Why should pronghorn form large groups, potentially increasing the level of food competition to its peak, at a time when food resources reach their nadir? I decided to investigate this question by finding out whether the rate of food competition really did increase in the large winter groups. I didn't want to just assume that it did, even though such a group-size effect had been reported in several other hoofed mammals. I decided to observe a few marked individuals, and to observe one individual per day during all the daylight hours.

Around the solstice, morning light does not become bright enough to see the colors of ear tags until about 8:30 A.M. Long purple shadows creep over the snow at 3:00 P.M. and the gloom descends at 4:30. So the working day was about half as long as a summer workday, but sitting motionless in the truck was not nearly as much fun. I bought a propane heater that kept our fingers limber enough to write in data sheets, but still my assistants and I were stiff and creaky at the end of a day. Mercifully, the data from two successive winters indicated exactly the same results, so I concluded with some relief that we had enough data. The upshot? The rate of aggressive interactions in winter was just the same as the rate in spring, summer, or fall. The types of aggressive interaction were just the same too: feeding, bedding, and simple displacements occurred in the same proportions as they did in other seasons. Pronghorn were still pushing each other around in winter groups, but the greater density of individuals did not create a higher rate of aggression. So somehow, pronghorn are able to avoid what has traditionally been seen as the cost of grouping. The rate of aggression is high, but it does not fluctuate with group size. Pronghorn are just as nasty to each other in small summer groups as in large winter groups.

At the beginning of January, each embryo weighs 1 and 1/4 ounce and is 4 and 1/2 inches long. January brings blizzards. The pronghorn recline with their rumps toward the howling wind and tuck their noses under a front leg. When they stand, they shake like a wet dog does, and snow flies in all directions. Ice balls form in the mane on the back of the neck. In the clear bitter cold after a blizzard, golden eagles hover over a herd, looking for a weak individual that they may be able to bring down. The pronghorn move to the lee side of wind-swept ridges, where the snow is shallow and the wind speed is reduced a little. On the high hard ground they may be chilled more quickly, but they are able to run fast if a coyote tries to get lucky or an eagle descends. In some years, there is a mid-January thaw. Daytime temperatures may reach 50°F, the snow melts, and for a moment the pronghorn have a greater choice of what to eat. The thaw lasts for only about a week and then winter closes in again. Snow piles up and fluctuating temperatures cause a hard crust to form. The narrow legs of speedsters punch through the crust and the animals plunge and struggle to make slow headway. The narrow legs are also poor snow shovels. As the animals continue to paw craters in search of plants to eat, the crust scrapes their legs and makes them bleed. Dominant individuals sometimes take over craters that other pronghorn have dug.

At the beginning of February, each embryo weighs 6 ounces and is 7 inches long. When deep snow and below-zero temperatures persist for several weeks, the pronghorn that were leanest at the solstice now run out of fat and begin to die. These individuals invariably are males. A really severe winter kills many adult males and some male fawns. It may seem odd that females, the sex with the greater energy expenditures now (those growing embryos), are less affected than the males. The disparity reveals the extent of the effort that males had to put into the rut. They entered winter with less fat than females because of all of the running, chasing, herding, and fighting they did in September.

At the beginning of March, the embryos weigh 18 ounces and are 9 inches long. In this part of Montana, on the west side of the continental divide, winter is almost over, but the season of death is not. The grassland plants will not begin any new growth for another 2 months, and most animals are dangerously near the end of their fat reserves. Moderating temperatures and snowmelt allow the pronghorn to go and to forage where they choose. The large groups begin to fragment, and older males begin to establish a solitary existence on the areas where they will defend harems next September.

At the beginning of April, each embryo weighs 2 and 1/2 pounds and is 13 inches long. Females now begin to feel the squeeze. From now until it is born, each fawn will almost quadruple its weight. The fawns are demanding more and more resources from the female now, at a time when those resources are dwindling fast. April truly is the cruelest month. The snow has melted and the ground has thawed, but nighttime temperatures still drop below freezing, and heavy clouds bring wet snow or freezing rain. The grassland is sodden, brown, and lifeless. The only plants to eat are those that grew last summer and their nutritional value is almost nil.

Toward the end of April, blackbirds begin to settle in cattails growing in the glacial potholes all over the valley. If the road is not too muddy, I drive south to look for smaller pronghorn groups and to see what birds are at Ravalli Ponds. In the reeds around the ponds the red-winged blackbird males have set up their territories and from high perches are proclaiming ownership again and again. A red-winged male has brilliant orange-red, almost day-glow epaulettes on the angle of each wing. The male cocks his wings out a bit, hunches forward, and utters a grating metallic sound not unlike that made by fingernails scraped on a blackboard. These displays are happening in wetlands all across North America. At Ravalli Ponds and elsewhere in the valley, a less common relative, the yellow-headed blackbird, is also present. Yellow-headed males are larger and more aggressive than red-winged males, and they often evict red-wings from choice habitats. Yellow-headed males also have the hunched-forward display, and their song is, if possible, even more bizarre than red-wing song. Here is one author's description of a male yellow-headed blackbird displaying to a female on his territory:

> Grasping a reed firmly in both fists, he leans forward, and, after premonitory gulps and gasps, he succeeds in pressing out a wail of de-

spairing agony which would do credit to a dying catamount. When you have recovered from the first shock, you strain the eyes in astonishment that a mere bird, and a bird in love at that, should give rise to such a cataclysmic sound. But he can do it again, and his neighbor across the way can do as well—or worse.[2]

I find it fascinating that blackbirds and meadowlarks are very close relatives (both in the family Icteridae), but have songs that evoke such disparate responses from human listeners.

The female blackbirds will arrive soon and each will settle on a territory where she will build a nest, incubate eggs, and forage for insects to feed nestlings. Some males attract several females and some attract one or none. A female may flit from one male to another for several days before she chooses one. Unlike pronghorn, however, the female blackbirds are evaluating males on the basis of the resources they offer, so the basis for comparing males is much more straightforward. Also, the trip from one male to the next is about 5 yards and takes about 2 seconds. A week or so after the blackbirds begin their raucous chorus I hear the first sweet meadowlark song, signaling the beginning of new life on the grassland. For the pronghorn, the season of starvation is over.

At the beginning of May, each embryo weighs 4 and 1/2 pounds and is 15 inches long. It will be twice that size when it emerges into the grassland in a month or less. May puts the greatest demand of gestation on the mother. But the mother's estrus was timed so that this final month and the beginning of lactation would coincide with the spring green-up. Now the grassland comes to life and the wonderful succession of flowering plants of the Palouse Prairie begins with spring beauty, prairie shooting star, glacier lily, and, in the creek bottoms, trillium. The 30 or so species of grasses send up new shoots and the Range turns emerald green. At the middle of the month, the first yellow balsamroot flowers begin to dot the grassland, and a week later there are big splashes of yellow, as if a huge brush has swept here and there over the green canvas. Sky-blue lupines and darker larkspur come up almost everywhere. The rarer bluish-purple camas emerge as single individuals or in small groups, and on drier slopes bitterroot display their pale pink blossoms.

At the peak of the balsamroot bloom the pronghorn groups are small again. Mature males are solitary on their chosen areas. Their new meat hooks are hard and sharp. The bachelor herds have formed, adding the clack-clack sound of sparring to the swelling grassland sounds that will continue unabated until next fall's first hard frost. The pronghorn females are now obviously pregnant— they waddle slightly when they walk, and they look ragged because they are starting to lose patches of winter hair. Small groups of females drift from place to place with little discernible goal. The food is so nutritious now and so abundant that any place is just as good as another.

Then on May 18, a female leaves her group. She walks slowly and carefully and she often stops to stand motionless, looking and listening. Two hours later she is completely alone. She is in a shallow bowl on a ridge that extends from Ravalli Ponds to the timber at Trisky Point. She reclines but does not chew her cud. Ten minutes later she stands up and her back arches upward in an exaggerated stretch. Her tail is flipped up. She circles in place and reclines. Suddenly there is a strong contraction of her uterus. Her neck stretches parallel to the ground and she groans. Over the next half hour the contractions come faster and faster and appear to become stronger. She stretches her neck to the side, lays her head on the ground, and emits a soft bleat. She stands up and circles, and I see a pair of pale hooves emerging from her vagina. She reclines again and has three more strong contractions. She stands and now the fawn's forelegs and head are out. The fawn looks dark—almost a chocolate brown. She has another contraction while standing and the fawn slips out and falls to the ground. The fall snaps the umbilical cord. She turns and begins to lick the fawn. The fawn has its head up and its ears flop loosely to the side. The mother licks rapidly and the fawn dries quickly in the afternoon sun. The mother reclines again and has four more contractions that expel the placenta. She gets up, noses the placenta, then grasps it with her teeth and eats it, bolting it down in 3 gulps. She resumes licking the fawn, which is now trying to stand. The fawn has almost ridiculously long legs and it wobbles and falls several times before it manages, shakily, to stay up. The mother continues to lick it as it blinks in the strong sunlight and reaches up for a nipple. The fawn nuzzles the mother's armpit, flank, and hind leg before it finds a nipple and begins to suck. The mother stands motionless, her back hunched up a bit to allow the fawn to stand directly beneath her. She cranes her neck down and sniffs and licks the fawn's anus. She turns away abruptly and the fawn is knocked off its feet when the next contraction hits.

In another half hour there are two small dark heads in the grass next to the resting mother. The mother stays down for almost another hour, resting and licking the fawns' heads, ears, and backs. She stands and both fawns struggle to their feet. There is a small streak of blood on the mother's white rump patch but otherwise no sign that she has just delivered two 8-pound fawns. The fawns suckle for almost 5 minutes while the mother stands patiently. She turns and licks both again and both assume the rump-up posture while the mother licks their bottoms. The mother reclines and the fawns recline beside her. An hour later, the mother stands and the fawns get up, now obviously stronger and hardly wobbling at all. The mother walks away slowly and the fawns follow, stumbling now and then but gamely keeping up. After a trip of about 500 yards, the mother stands and looks in all directions while the fawns stand beneath her. I can barely see them—they could almost be a shadow below the mother.

Now the mother steps away and gives the fawns a signal. They turn and walk away from her to find their first hiding place. The mother walks away from the fawns but whips her head around as the first then the second fawn reclines. She knows exactly where they are and she is now engaged in a battle to keep them alive. If they live for 3 weeks they will be among the first fawns in a summer group, and they will become dominant over all other fawns. If they survive to 50 days, they will become adults. That seems like such a

A mother and her first fawn 15 minutes after birth. The second fawn is not born yet. The showy flowers are arrowleaf balsamroot.

short time. But when every minute can bring a hunting coyote or a golden eagle, 50 days becomes an almost endless series of crises.

As I drive home in the early evening after watching the birth, the air is soft and slanting light picks out several waterfalls that appear as silver lines etched in the dark sides of the Mission Range. The mountains still carry a heavy cloak of snow and I imagine how those waterfalls must be roaring. I stop above the Ravalli Ponds and get out to sit on a knoll to look for ruddy ducks. I look carefully before I sit down because the rattlesnakes are out now. I sit in the new grass and study the pond and watch a beetle feeding in a balsamroot flower next to me. I wonder what surprises the summer will bring and I start to think about which males are likely to be successful this fall. Then I hear it—a tiny sound from high up that almost seems to start from inside my head, but then swells to fill the evening air:

woo woo woo woo woo woo woo WOO WOO woo woo woo woo...

13 *THE FLOOR OF THE SKY*

Elsewhere the sky is the roof of the world; but here the earth was the floor of the sky.

Willa Cather, *Death Comes for the Archbishop*

13 The University of Idaho, were I work when I am not on the Bison Range, is in Moscow, a little town in the northern third of the state and right on the border with the state of Washington. I can drive to the Bison Range in 4 hours but the topography around Moscow is very different from the topography of the Flathead Valley. Moscow lies just west of the Clearwater and Bitterroot Mountains in a landscape of big choppy hills that from above look like a stormy sea frozen in motion or like a desert of sand dunes. The dunes simile is more precise, because these hills arose from fine soil particles that blew here. Geologists call this loess topography. The loess here came from lake-bottom sediments in central Washington. When the lakes dried, the sediments were carried by the prevailing southwest wind as far as the base of the mountains, forming a 5,000-square-mile region of dirt dunes. After the sediments blew in, they were colonized by native short grass prairie species that stabilized the hills and enriched the soil.

By the time the farmers arrived in the 1860s, this region had acquired the name Palouse. The farmers found rich topsoil that was over 6 feet deep in places and they quickly plowed under the prairie. Today the principal crop is soft white wheat and the largest surviving remnant of Palouse Prairie is 225 miles away, on the Bison Range. The Bison Range was established in 1908 to rescue bison from the brink of extinction but inadvertently the conservation effort on behalf of a single species became an act that preserved an entire native plant community. The Bison Range still is best known for its big grazers, but it has far greater biological significance as the largest intact remnant of Palouse Prairie.

Some of my happiest moments on the Range occur when, to get to a fawn or to get to a vantage point where I will watch a harem, I need to hike for a mile or so through the heart of the grassland. The grassland is like a garden, but it is more complex than any human garden and, for me, more satisfying. Grasses whisper against my pant legs. The soil is springy. Flowers of the season, and always some species that I haven't seen before, are everywhere. A vole runs across my shoe. To my left, a short-eared owl hovers with strange, slow, butterfly-like wing beats just above the grass tops. A bison herd 2 miles away is a dark smudge against the base of Antelope Ridge. Vesper sparrows flush in front of me and fly low to the ground before they drop back out of sight. I stop to look at an elk antler that a porcupine has been gnawing. Near Indian Springs and its little grove of aspens, I surprise a mother black bear that is teaching her cub how to dig up the corms of spring beauties. Fortunately, they see me before I get too close and they quickly walk away, the mother in the lead. As I admire their intense, velvety-black coats, the bears startle an elk mother that has a calf hiding somewhere nearby. The mother prances away, her head held back and her feet coming up high in that goofy gait that elk

Grassland: the Alexander Basin of the National Bison Range on a windy morning in May.

use when they're alarmed. Meadowlarks are burbling everywhere, filling the air with melody.

The grassland is satisfying because it is complex, but there's more to it than that. Rainforests and reefs are complex, but they don't seem to evoke the same kind of feeling. In both of these environments I admire the biological richness that surrounds me—I won't deny that any kind of untouched natural diversity is a wondrous thing. But in both reef and rainforest I feel out of my element, almost as if I am trespassing. By contrast, when I walk through the grassland on the Range, and especially when I walk where there are a few scattered trees, I have the deeply relaxing sensation that I am home. And that is not because I grew up in grassland: When Charlie and I were lobbying for that first TV our family lived in a small piedmont town in Virginia and later we moved to suburban Baltimore-Washington. Also, I don't think that I feel at home in the grassland just because I've worked there for 20 years. I remember having that same at-home sensation during the first field season and I rediscover it with delight each spring. Even though Willa Cather lived her adult life in New York City, she never stopped writing about the feeling that is associated with the prairie. Many biologists have noted that a deep emotional response to grassland makes sense from an evolutionary perspective, because grassland dotted with trees is the environment where our species came into being. Only in the very recent past did our species move out of the African savanna to colonize the rest of the world. Like the pronghorn that act as if they feel the presence of the hyenas and big cats in their past, we continue to show our evolved habitat preferences in our emotional response to grassland.

Unfortunately, most people never have the opportunity to walk through natural grassland. But to make up for the loss, people recreate grassland in microcosm as suburban lawns and on a slightly larger scale in urban parks. The pull of grassland is also revealed by a huge literature on the ecology, history, economics, and creative response to the prairie, by the dozens of prairie restoration groups, and by the surprisingly rapid, widespread, and diverse response to the Buffalo Commons, a proposal for land use in the Great Plains.[1]

The evolutionary biologist Richard Dawkins noted that thoughts and ideas, because of language, reproduce from one mind to another, and because of the nature of minds, also mutate. Dawkins coined the term meme to describe the notion of evolving ideas, and noted that meme evolution can occur much more quickly than gene evolution. A good example of a meme is the Buffalo Commons, a notion conceived by Frank and Deborah Popper of Rutgers University.

In 1987, the Poppers proposed that a huge swath of the Great Plains, from west of the 98th meridian to the Rockies, should be allowed with the aid of creative public policy to revert to grassland where vast bison herds could roam. The Poppers argued that this would be the wisest land use policy and one far better than the unrealistic system of government subsidies that led humans to impose agriculture on a landscape that is ill suited to it.[2] The idea reproduced with the speed of a virus, and it mutated and evolved.[3] One of its descendants is the Great Plains Restoration Council, a group devoted to large-scale prairie restoration and in the near term to the creation of a million-acre reserve. I think that the Buffalo Commons meme had such great force because of the deep instinctive pull of grassland on the human spirit.

There are many reasons for restorative conservation in the Great Plains. The reasons range from the pragmatic, to the appeal to history, to native rights, to preservation of particular species. A grasslands preserve would certainly be a good thing for bison. When they were snatched from the brink of extinction in the late nineteenth century, bison were seeded into a few small, scattered reserves. The number of reserves has grown, but genetic interchange is slight, and some populations now show signs of genetic incompatibility with each other.[4] Moreover, genes from domestic cows are not uncommon in ostensibly pure stocks of bison.[5] Female (but not male) hybrids of domestic cattle and bison are fertile, so they are an avenue by which genes from domestic cattle can be introduced into the bison genome.[6] Bison are no longer on the sinking ship but they are in the lifeboats. The rescue won't be complete until a large, free-ranging wild population is established.

To me, however, the most powerful reason for a Great Plains reserve is an aesthetic one. Collectively, at our best we strive for beauty and because of who we are, because of our origins, a huge Great Plains reserve would be a beautiful thing. A lesser but still important reason is that in such a reserve we would lead by example. It is hypocritical to exhort the Brazilians to conserve their rainforest after we have already destroyed the grassland ecosystem that occupied half the continent when we found it. A large-scale grassland restoration project would give us some moral authority when we seek conservation abroad.

I must admit that I also like the idea because it would mean a better home for pronghorn, currently pushed by agriculture into marginal habitats—the high sagebrush deserts of the West. I would love to return the speedsters to their evolutionary home, the Floor of the Sky. Imagine a huge national reserve where anyone could see what caused Lewis and Clark to write with such enthusiasm in their journals—the sea of grass and flowers dotted with massive herds of bison, accompanied by the dainty speedsters and by great herds of elk. Grizzly bears and wolves would patrol the margins of the herds and coyotes would at last be reduced to their proper place. The song of the meadowlarks would pervade the prairie and near water the spring air would ring with the eerie tremolos of snipe.

NOTES

1. ANATOMY OF A SPEEDSTER

1. J. Murie, "Notes on the anatomy of the prongbuck, *Antilocapra americana*," *Proceedings of the Zoological Society of London*, 36 (1870): 353.
2. J. J. Audubon and J. Bachman, *The Quadrupeds of North America*, vol. 2 (New York: V. G. Audubon, 1851), p. 198.
3. T. M. McKean and B. Walker, "Comparison of selected cardiopulmonary parameters between the pronghorn and the goat," *Respiratory Physiology*, 21 (1974): 365–370.
4. S. L. Lindstedt, J. F. Hokanson, D. J. Wells, S. D. Swain, H. Hoppeler, and V. Navarro, "Running energetics in the pronghorn antelope," *Nature*, 353 (1991): 748–750.
5. R. E. Bullock, "An analysis of locomotor body movements in pronghorn antelope," *Canadian Journal of Zoology*, 60 (1982): 1871–1880.
6. C. Carlton and T. McKean, "The carotid and orbital retia of the pronghorn, deer and elk," *Anatomical Record*, 189 (1977): 91–107.
7. B. Kurten and E. Anderson, *Pleistocene Mammals of North America* (New York: Columbia University Press, 1980).
8. M. Antón and A. Turner, *The Big Cats and Their Fossil Relatives: An Illus-*

trated Guide to Their Evolution and Natural History (New York: Columbia University Press, 1997).

9. J. A. Byers, *American Pronghorn: Social Adaptations and the Ghosts of Predators Past* (Chicago: University of Chicago Press, 1997).

2. SPRING AND THE SOUNDS OF SNIPES

1. First proposed in C. R. Darwin, *On the Origin of Species by Means of Natural Selection, or the Preservation of Favored Races in the Struggle for Life* (London: Murray, 1859), and later elaborated upon in C. R. Darwin, *The Descent of Man, and Selection in Relation to Sex* (London: Murray, 1871).

2. J. A. Byers and K. Z. Byers, "Do pronghorn mothers reveal the locations of their hidden fawns?" *Behavioral Ecology and Sociobiology,* 13 (1983): 147–156.

3. J. A. Byers, "Mortality risk to pronghorns from handling," *Journal of Mammalogy,* 78 (1997): 894–899.

4. J. Muir to K. Hittell, April 1865, reprinted in the *John Muir Newsletter,* 1, no. 4 (1991): 3–4.

5. R. O. Prum, R. H. Torres, S. Williamson, and J. Dyck, "Coherent light scattering by blue bird feather barbs," *Nature,* 396 (1998): 28–29.

3. FIRST FIELD SEASON

1. W. F. Wood, "2-pyrrolidinone, a putative alerting pheromone from rump glands of pronghorn, *Antilocapra americana,*" *Biochemical Systematics and Ecology,* 30 (2002): 361–363.

2. C. Barlow, *The Autobiography of Charles Darwi,n 1809–1882* (New York: W. W. Norton, 1969), p. 70.

3. F. Fournier and M. Festa-Bianchet, "Social dominance in adult female mountain goats," *Animal Behaviour*, 49 (1995): 1449–1459.

4. THE ADULT BULLIES

1. Starlings are impressive vocal mimics and in nature imitate the songs and calls of many other bird species. One starling that lived with humans repeated the phrase "Does Hammacher Schlemmer have a toll-free number?" after hearing it once: M. P. West and A. P. King, "Mozart's starling," *American Scientist*, 78 (1990): 106–114.
2. J. Altmann, "Observational study of behavior: sampling methods," *Behaviour*, 49 (1974): 227–267.
3. C. FitzGibbon, "A cost to individuals with reduced vigilance in groups of Thomson's gazelles hunted by cheetahs," *Animal Behaviour*, 37 (1989): 508–510.

5. MILK POLITICS

1. For a lucid explication of the view that organisms are the complicated robots built and run to serve the replicating molecules they house, see R. Dawkins, *The Selfish Gene* (New York: Oxford University Press, 1976), and R. Dawkins, *The Extended Phenotype: The Gene as the Unit of Selection* (Oxford: W. H. Freeman, 1982).
2. G. J. Vermeij, *Evolution and Escalation: An Ecological History of Life* (Princeton: Princeton University Press, 1987).
3. G. J. Vermeij, "Inequality and the directionality of history," *The American Naturalist*, 153 (1999): 243–253.
4. A. F. Bennett, "Activity metabolism of the lower vertebrates," *Annual Review of Physiology*, 400 (1978): 447–469.

5. O. T. Oftedal, "Milk composition, milk yield and energy output at peak lactation: a comparative review," in M. Peaker, R. G. Vernon and C. H. Knight, ed., *Physiological Strategies in Lactation, Symposia of the Zoological Society of London,* 51 (1984): 33–85.

6. J. R Powell, M. E. J. Barratt and P. Porter, "The acquisition of immunity by the neonate," in D. B Crighton, ed., *Immunological Aspects of Reproduction in Mammals* (London: Butterworths, 1984) pp. 265–290.

7. R. L. Trivers, "Parent-offspring conflict," *American Zoologist,* 14 (1974): 249–264.

6. LITTLE SPEEDSTERS

1. M. N. Miller and J. A. Byers, "Energetic cost of locomotor play in pronghorn fawns," *Animal Behaviour,* 41 (1991): 1007–1013.

2. J. A. Byers and C. B. Walker, "Refining the motor training hypothesis for the evolution of play," *The American Naturalist,* 146 (1995): 25–40.

7. COLUMNS OF DUST

1. J. Berger and C. Cunningham, *Bison: Mating and Conservation in Small Populations* (New York: Columbia University Press, 1994), pp. 119–128.

2. Ibid., pp. 170–176.

8. BACHELOR WORKOUT

1. W. D. Hamilton, "Geometry for the selfish herd," *Journal of Theoretical Biology,* 31 (1971): 295–311.

9. THE TURNING YEAR

1. G. E Belovsky and J. B Slade, "Insect herbivory accelerates nutrient cycling and increases plant production," *Proceedings of the National Academy of Science,* 97 (2000): 14412–14417.
2. J. A. Byers and J. T. Hogg, "Environmental effects on prenatal growth rate in pronghorn and bighorn: further evidence for energy constraint on sex-biased maternal expenditure," *Behavioral Ecology,* 6 (1995): 451–457.
3. Ibid.

10. MAKING NEXT YEAR'S FAWNS

1. J. A. Byers, J. D. Moodie, and N. Hall, "Pronghorn females choose vigorous mates," *Animal Behaviour,* 47 (1994): 33–43.

11. AFTER THE EQUINOX

1. G. E. Moulton, ed., *The Journals of the Lewis and Clark Expedition,* vol. 7 (Lincoln: University of Nebraska Press, 1983), entry for Tuesday, 27 May 1806.
2. J. T. Hogg, "Mating in bighorn sheep: multiple creative male strategies," *Science,* 225 (1984): 526–529.
3. J. T. Hogg and S. H. Forbes, "Mating in bighorn sheep: frequent male reproduction via a high-risk 'unconventional' tactic," *Behavioral Ecology and Sociobiology,* 41 (1997): 33–48.
4. We know what happens inside a female pronghorn's uterus because of the Ph.D. dissertation research of Bart O'Gara, who subsequently

was leader of the Cooperative Wildlife Research Unit at the University of Montana for many years. In the early 1960s, Bart shot 104 pregnant pronghorn females at different stages of gestation, and carefully described the contents of the uterus. See B. O'Gara, "Unique aspects of reproduction in female pronghorn," *American Journal of Anatomy,* 125 (1969): 217–231.

5. E. C. Birney and D. Baird, "Why do some mammals polyovulate to produce a litter of two? *The American Naturalist,* 126 (1985): 136–140.

1 2 . A F T E R   T H E   S O L S T I C E

1. T. J. Ryder and L. L. Irwin, "Winter habitat relationships of pronghorns in south-central Wyoming, *Journal of Wildlife Management,* 51 (1987): 79–85.
2. W. L. Dawson, *The Birds of California,* vol. 1 (San Diego: South Moulton Co., 1923), p. 126.

1 3 .   T H E   F L O O R   O F   T H E   S K Y

1. For some examples, see E. Callenbach, *Bring Back the Buffalo! A Sustainable Future for America's Great Plains* (Washington, D.C.: Island Press, 1996), and D. O'Brien, *Buffalo for the Broken Heart: Restoring Life to a Black Hills Ranch* (New York: Random House, 2001).
2. J. M. Campbell, *Magnificent Failure: A Portrait of the Western Homestead Era* (Stanford: Stanford University Press, 2001).
3. To read about the range of public responses to the Poppers' proposal, see A. Matthews, *Where the Buffalo Roam* (New York: Grove Weidenfeld, 1992).

4. J. Berger and C. Cunningham. *Bison: Mating and Conservation in Small Populations* (New York: Columbia University Press, 1994).

5. T. J. Ward, J. P. Bielawski, S. K. Davis, J. W. Templeton, and J. N. Derr, "Identification of domestic cattle hybrids in wild cattle and bison species: a general approach using mtDNA markers and the parametric bootstrap," *Animal Conservation,* 2 (1999): 51–57.

6. T. J. Ward, L. C. Skow, D. S. Gallagher, R. D. Schnabel, C. A. Hall, C. E. Kolenda, S. K. Davis, J. F. Taylor, and J. N. Derr, "Differential introgression of uniparentally inherited markers in bison populations with hybrid ancestries," *Animal Genetics,* 32 (2001): 89–91.

ACKNOWLEDGMENTS

Many assistants and several graduate students worked, and are still working, with me on the Bison Range. They enrich me and the project and I thank them for their dedication and especially for their enthusiasm.

I thank the U.S. Fish and Wildlife Service and the National Bison Range manager, Dave Wiseman, for permission to do my research at this unique site. Managers of the Bison Range and other similar facilities face many conflicting demands for public use and I am grateful for Dave's consistent support of my research.

Russell Tucker of the Washington State University College of Veterinary Medicine graciously made an X-ray of the smelly pronghorn leg I brought to him. Dan Bukvich and Michael Locke of the University of Idaho Lionel Hampton School of Music transcribed a sonogram of meadowlark song into musical notation.

My research from 1981 to the present was supported by grants from the National Institutes of Health (17029 and 22606), the National Geographic Society (4759–92), and the National Science Foundation (IBN 9808377 and IBN 0097115).

At Harvard University Press, I thank Michael Fisher for astute guidance and Nancy Clemente for a truly graceful editing hand.

Finally, I want to offer my thanks to the pronghorn of the Bison Range for tolerating my presence as they have for many years. They conducted all of the important business of their lives often within 20 feet of my truck and thus allowed me a privileged glimpse into another society. They didn't have to remain close when they did—they could easily have moved to anywhere in the 30 square miles of the Range to avoid me, but often they stayed close; sometimes a group of fawns, bright and inquisitive, walked to the truck and stared at me. I also learned that even open-country animals occasionally want some privacy and I adopted the term N.A.V. (Not Available for Viewing) to indicate a group that I should not try to pursue. Slowly, I learned some pronghorn manners.

INDEX